Radio Interfaces in the Internet of Things Systems

Kamil Staniec

Radio Interfaces in the Internet of Things Systems

Performance studies

Springer

Kamil Staniec
Department of Telecommunications
and Teleinformatics Faculty of Electronics
Wrocław University of Science and Technology
Wrocław, Poland

ISBN 978-3-030-44845-5 ISBN 978-3-030-44846-2 (eBook)
https://doi.org/10.1007/978-3-030-44846-2

This Springer imprint is published by the registered company Springer Nature Switzerland AG
The registered company address is: Gewerbestrasse 11, 6330 Cham, Switzerland

To my beloved wife, Anna, and my daughters, Oliwia and Jagoda

Preface

This book has grown out of my teaching and research experience at the Wrocław University of Science and Technology (WUST). My scientific interests began with the radio waves propagation modelling, then moved on to studies on wireless short-range systems to finally arrive at the electromagnetic (EM) compatibility in wireless sensor networks and Internet of Things (IoT) systems. Such an evolution has yielded lots of references in the book concerning the issue of IoT EM environments that is prone to become all the more challenging in the years to come, due to inadvertent jamming that will arise as increasingly many IoT networks will coexist at the same geographical area, sharing the same narrow frequency bands. Therefore, apart from conceptual considerations (such as the market analysis, state-of-the-art and forecasted IoT deployment, etc.) and descriptions of operational principles that make IoT technologies so unique in the wireless ecosystem, the book pays a great deal of attention to the problem of EM cybersecurity, often underestimated in similar works.

Readers who may want to reach for this book (in particular its Chaps. 1, 2 and 5) are associated with all industrial sectors related to handling the machine-type-traffic (MTC) or machine-to-machine (M2M) communications, mainly dealing with the provision of utilities to end-consumers, for example, municipal waterworks and energy and gas providers. Some may have already had experience with gathering readings from their meters (detectors, gauges, etc.) wirelessly using short-range systems such as ZigBee or Wireless M-Bus that require regular touring for acquiring data near locations where the devices are installed. The book may therefore draw their interest to long-range solutions (known as IoT systems) by demonstrating that these enable eliminating this 'touring requirement' and make it possible to send data from utility devices directly to the company's premises and back (!), since IoT solutions allow for bi-directional communications, thus their owners may have a full control on end devices by, for instance, switching them on/off, updating their software, modifying their reporting schedule, uploading new policies, etc.

Another potentially interested group are designers of IoT systems networks who offer their services to the aforementioned utilities companies, who may find Chaps. 2, 5 and 7 particularly interesting. The book provides them with quantitative analysis of a wide range of Low-Throughput Networks (LTN) IoT systems that may come in handy with calculating some of the most crucial network parameters, i.e. coverage and capacity. Both these aspects were subject to intensive investigations which results, both theoretical and measurement-based, are exhaustively discussed throughout the book with some of the findings published for the first time (like Chap. 6 and Sects. 7.2 and 7.3). All the measurements were carried out in a unique, specialized suite of laboratories for electromagnetic compatibility at WUST, which, due to their excellent EM isolation, allowed for achieving, among others, ultra-high dynamic operational ranges for high-sensitivity IoT systems, obtained in anechoic and semi-anechoic chambers, as well as heavily-reflective radio channels, attainable in the reverberation chamber.

Chapters 1, 2, 3, 4, 5, 8 and 11 may turn out helpful for university students in telecommunication engineering, from the standpoint of understating the operation of various radio interfaces used in IoT systems (though not only), explained with avoidance of a scientific language, focusing on clarifying how particular multiple-access or radio interfaces affect the energetic consumption of end-devices or determine their sensitivity and/or vulnerability to EM interference. Chapter 3 teaches methods for calculating pathloss in different environments, accounting for the Line-of-Sight probability, types of buildings and a degree of required resultant precision. It was my intention to make the entire scope of the book (with Chaps. 6 and 7 only for particularly motivated students) a prerequisite for subjects such as: 'Internet of Things', 'Sensor Networks' or 'Wireless Networks' and partially recommended for other classes such as 'the Radiowaves Propagation' (Chap. 3) or 'Transmission Media' (Chap. 3 and partially – with a teacher's guidance – also elements of Chaps. 6 and 7). All of these subjects are taught as a part of curriculum in both bachelor's and master's programmes at WUST.

I believe the aforementioned readers groups will draw tangible benefits from the book, some of which, depending on the audience type, include:

- Companies wishing to switch to using IoT technologies for handling their sensor traffic: a complete guide to selecting an optimal IoT technology to suit their sensor traffic needs along with engineering formulas for calculating coverage, capacity and data rates. They will be provided with a thorough, easy to comprehend explanation of benefits resulting from replacing traditional cellular technologies with IoT systems, written from a down-to-earth, engineering viewpoint (that I myself represent), omitting commercial-sounding clichés and gimmicks. All of that is served in the form of a practical guide, helping to figure out the most optimal choice for a given use-case.
- IoT hardware producers: they can find Chaps. 6 and 7 particularly handy in their pre-launch processes of validating their devices performance in harsh target environments wherein their products may have to operate. The chapters provide a full methodology for reproducing radio channels of various interference and multipath fading types in artificial conditions allowing for both, generation of fully controlled physical disturbing phenomena as well as obtaining repeatable results. Chapter 8 provides validation results of performance test carried out with this methodology on selected specimens of real IoT systems.

- Students of electronic/communication engineering: a comprehensive tutorial on how different methods for spectrum spreading, medium access control and physical signals transmission, work in practice. Academic talk is limited to only the two final chapters, leaving most of the book to non-scientists. This audience is given a detailed yet easy-to-follow guide on techniques used in IoT (and not only) radio interfaces, with explanations – not merely descriptions – of mechanisms behind them as well as a tutorial on pathloss calculation and spectrum-related issues.
- IoT network designers and IoT hardware developers: formulas for data rate, capacity and coverage calculation, tables informing of available frequency spectra and limitations imposed on them (i.e. the radiated power, duty cycle, etc.) by international institutions such as ETSI, ITU, ERC, etc. The key feature presented to these readers consists in providing an innovative step-by-step methodology for measuring IoT devices performance under variable-quality propagation conditions obtained in controlled environments yielding repeatable and reliable outcomes.

Wrocław, Poland Kamil Staniec

Contents

1 Internet of Things: Introduction ... 1
1.1 IoT: Specificity, Goals and Hopes ... 1
1.2 Internet of Things: Definitions and Systematics ... 6
1.2.1 The Internet of Things (IoT) Concept ... 6
1.2.2 Technical Description of IoT ... 7
1.2.3 IoT Basic Characteristics and High-Level Requirements ... 10
1.2.4 The IoT Reference Model ... 11
1.3 IoT: Risks and Concerns ... 14
1.4 New Concepts and Development Forecasts ... 17
1.5 The IoT Associated Concepts and Derivatives ... 18
1.5.1 The Ambient Intelligence (AmI) ... 19
1.5.2 Edge Intelligence and Cloud/Fog Calculations ... 23
1.5.3 Capillary Networks and Media Gateways ... 28
1.5.4 A Statistical Traffic Model for the Machine-Type Communications ... 31

2 IoT Networks Standardization and Legal Regulations ... 33
2.1 The Smart Metering Concept ... 33
2.2 The IEEE P2413 Standard ... 36
2.3 Standardization According to ETSI: LTN ... 38
2.3.1 Application Scenarios of the LTN (ETSI GS LTN 001) ... 41
2.3.2 A Functional Architecture of the LTN (ETSI GS LTN 002) ... 50
2.4 Spectral Issues ... 56
2.5 Medium Access Techniques in the IoT ... 58
2.5.1 The CCA Mechanism ... 58
2.5.2 The AFA Technique ... 59
2.5.3 The LBT Technique of Polite Medium Access ... 59

3 Propagation Models for LTN Systems ... 61
3.1 The Hata Model ... 61
3.1.1 Urban Areas ... 62
3.1.2 Suburban Areas ... 62
3.1.3 Open Areas ... 62
3.2 WINNER+ Model ... 63
3.3 A Simple 3GPP Model ... 63
3.4 An Extended 3GPP Model ... 63
3.4.1 Pathloss ... 63
3.4.2 LOS Conditions Probability ... 70
3.5 Calculation of Additional Losses ... 70
3.5.1 Losses on Scattering Obstacles ... 70
3.5.2 Building Penetration Losses: A 3GPP Model ... 73
3.5.3 Building Penetration Losses: WINNER+ and ITU-R Models ... 73

4 Radio Interfaces in LTN Networks ... 79
4.1 Spread Spectrum Techniques Used in IoT Systems ... 80
4.1.1 Direct Sequence Spread Spectrum (DSSS) ... 80
4.1.2 Chirp Spread Spectrum (CSS) ... 82
4.2 IoT Systems with Spread Spectrum (OSSS): General Characteristics ... 85
4.3 Ultra-Narrowband Transmission: UNB Systems ... 87
4.3.1 UNB Systems Uplink (UL) Radio Interface ... 88
4.3.2 UNB Systems Downlink (DL) Radio Interface ... 90
4.3.3 RFTDMA Multiple Access Mechanism in UNB Systems ... 91

5 The Internet of Things Narrow-Band LPWAN/UNB Systems ... 93
5.1 LoRaWAN (LoRa): Architecture, Radio Interface and Device Classes ... 93
5.1.1 LoRaWAN: System Architecture ... 94
5.1.2 LoRaWAN: Device Classes ... 95
5.1.3 LoRa: Frame Structure, the Most Important Commands ... 97
5.1.4 Activation of the End Device ... 100
5.1.5 LoRa Physical Layer ... 100
5.2 Weightless (-P): Architecture, Radio Interface and Device Classes ... 110
5.3 SigFox: Architecture, Radio Interface and Device Classes ... 114

6 Performance Measurements Methodology for LTN IoT Systems ... 119
6.1 Electromagnetic Interference in Internet of Things Networks ... 119
6.2 Measurements of Immunity to Noise and Interference ... 121
6.3 Measurements of Immunity to Extremely Multipath Propagation ... 129

7 LPWAN/UNB IoT Systems Performance Investigations 137
7.1 LoRa System 137
7.1.1 Immunity to Noise and Interference 137
7.1.2 Immunity to the Multipath Propagation. 139
7.1.3 Summary 140
7.2 Weightless(-P) System 142
7.2.1 Immunity to Noise and Interference 142
7.2.2 Immunity to the Multipath Propagation. 144
7.2.3 Summary 146
7.3 SigFox System 147
7.3.1 Immunity to Noise and Interference 147
7.3.2 Immunity to the Multipath Propagation. 148
7.3.3 Summary 149

8 The Cellular Internet of Things Systems (CIoT) 151
8.1 Functional Assumptions for CIoT Systems 153
8.2 Energy Efficiency in CIoT Systems. 155
8.2.1 Power Saving Mode (PSM) 156
8.2.2 Extended Discontinuous Reception Mode (eDRX). 157
8.3 NB-IoT: Narrowband IoT 160
8.4 LTE-M(TC): LTE-Machine (Type Communication) 170
8.5 EC-GSM IoT 174

Annexes 179

Bibliography 183

Index. 191

Acronyms

AFA	Adaptive Frequency Agility
AmI	Ambient Intelligence
API	Application Programming Interface
ARF	Automatic Rate Fallback
AWGN	Additive White Gaussian Noise
BO	Back Off
BPSK	Binary Phase Shift Keying
BP	Base Station
BW	Bandwidth [Hz]
BW_{coh}	Coherence Bandwidth [Hz]
C_{b_agr}	Aggregated capacity of a LoRa channel of width BW [b/s]
CDMA	Code Division Multiple Access
C-eDRX	Connected-Extended Discontinuous Reception
CIoT	Cellular Internet of Things
CNR	Carrier-to-Noise Ratio
CNIR	Carrier-to-Noise-and-Interference Ratio
CSS	Chirp Spread Spectrum
CCA	Clear Channel Assessment
CR	Code Rate
DC	Duty Cycle
DL	Downlink
DRX	Discontinuous Reception
DSSS	Direct Sequence Spread Spectrum
EaaS	Education as a Service
EC-GSM	Extended Coverage GSM
EC-PACCH	Extended Coverage Packet Associated Control CHannel
EC-PDTCH	Extended Coverage Packet Data Traffic CHannel
EC-SCH	Extended Coverage Synchronization CHannel
EC-BCCH	Extended Coverage Broadcast Control CHannel
EC-CCCH/D	Extended Coverage Common Control Channel/Downlink

EC-CCCH/U	Extended Coverage Common Control Channel/Uplink
ED	End Device
eDRX	Extended Discontinuous Reception
eMBB	Enhanced Mobile Broadband
eNB	Evolved Node-B
ERP	Equivalent Radiated Power
ETSI	European Telecommunications Standards Institute
FDMA	Frequency Division Multiple Access
FEC	Forward Error Correction
FFT	Fast Fourier Transform
FHSS	Frequency-Hopping Spread Spectrum
FSK	Frequency-Shift Keying
G_p	Processing Gain
GPS	Global Positioning System
GPRS	General Packet Radio Service
HARQ	Hybrid Automatic Repeat Request
HFN	Hyper Frame Number
IaaS	Infrastructure as a Service
ICT	Information and Communications Technology
I-eDRX	Idle-eDRX
ISD	Inter-site Distance
ISI	Intersymbol Interferences
IoT	Internet of Things
ISM	Industrial, Scientific, Medical
IRR	Infrared Reflective
LAP	LTN Access Point
LBT	Listen Before Talk
LEP	LTN End Point
LPWAN	Low-Power Wide-Area Network
LTE-M(TC)	LTE-Machine-Type Communications
LTN	Low-Throughput Network
M2M	Machine-to-Machine
MAC	Medium Access Control
MAR	Mobile Autonomous Reporting
MCS	Modulation and Coding Scheme
MIB	Master Information Block
mMTC	Massive Machine-Type Communications
MME	Mobility Management Entity
MTC	Machine-Type Communications
OM	Operational Mode
MS	Mobile Station
MTD	Machine-Type Device

NB-IoT	Narrowband IoT
NF	Noise Factor [dB]
NO	Network Operations
N_{p_h}	Number of symbols composing the packet payload and header in LoRa system
N_{rep}	Number of repetitions
PaaS	Platform as a Service
PAYG	Pay As You Go
PBCH	Physical Broadcast Channel
PCH	Paging Channel
PDP	Power Delay Profile
PER	Packet Error Rate
PLL	Phased-Locked Loop
PMA	Polite Medium Access
P_{min}	Sensitivity
PO	Power Amplifier
PRB	Physical Resource Block
P-RNTI	Paging Radio Network Temporary Identifier
PSD	Power Spectral Density
PSM	Power Saving Mode
PSS	Primary Synchronization Signal
QPSK	Quadrature Phase Shift Keying
R	Code rate
RAU	Routing Area Update
R_b	Bit rate
RB	Resource Block
R_c	Chip Rate [c/s] or [chip/s]
RFID	Radio Frequency Identification
RRC	Radio Resource Control
R_s	Symbol Rate [sym/s]
SaaS	Software as a Service
SF	Spreading Factor
SFN	System Frame Number
SGSN	Serving Gateway Support Node
SIB	System Information Block
SISO	Single Input Single Output
SMS	Short Message Service
SNR	Signal-to-Noise Ratio
SNIR	Signal-to-Noise-and-Interference Ratio
SSS	Secondary Synchronization Signal
TAU	Tracking Area Update
TDMA	Time Division Multiple Access

TDM	Time Division Multiplexing
T_M	Time needed to transmit N_{p_h} symbols
T_p	Time needed to transmit a whole packet
T_{pre}	Time needed to transmit a preamble
T_s	Time needed to transmit a symbol
UE	User Equipment
UL	Uplink
UNB	Ultra-Narrowband
URLLC	Ultra-Reliable and Low Latency Communications
V2I	Vehicle-to-Infrastructure
V2V	Vehicle-to-Vehicle
Wi-Fi	Wireless Fidelity
WLAN	Wireless Local Area Network
WSN	Wireless Sensor Network

Chapter 1
Internet of Things: Introduction

1.1 IoT: Specificity, Goals and Hopes

In the era of matured Internet, it has been noticed that sensors, gauges and detectors also constitute a kind of 'network community' participating in network traffic, and generating not less important data, compared to that produced by man, and often strategic in maintaining quality of life, monitoring industrial processes or natural environment. These devices or, more generally, objects are capable of communicating either with men (operators) or with other devices. In the latter case, they are sometimes referred to in the literature as MTDs, i.e. machine-type devices, which do not require human intervention in data exchange. Likewise, the traffic generated by them is termed as the machine-type communications (MTC), and the process of communication between them is called M2M (*machine-to-machine*) and is one of the paradigms of the so-called Internet of Things (IoT) which is a branch of our global network wherein MTD devices are partners equivalent to their 'human' counterparts in terms of significance of telecommunications traffic generated by the MTD devices, increasingly more necessary to maintain seamless functioning of our technological civilization.

The term 'Internet of Things' was first put forward during one of the Procter & Gamble company presentations by Kevin Ashton in 1999 [1], although the author himself currently promotes another term, i.e. 'Internet for Things' [2]. It was initially anticipated that RFID (radio frequency identification) would be the technology used to implement the IoT tasks. It was expected to be a tool responsible for executing a variety of tasks assigned by computers to objects (things), such as actuators, counters, controllers, etc. [3–7]. The turn of 2008/2009 may be assumed to be an informal date of reaffirming the existence of the Internet of Things, when (according to Cisco's own research described in [8]) the quotient of 'objects' communicating with people in the network exceeded the value of '1', amounting to '1.84' in 2010, while in 2003 it was estimated at '0.08'.

K. Staniec, *Radio Interfaces in the Internet of Things Systems*,
https://doi.org/10.1007/978-3-030-44846-2_1

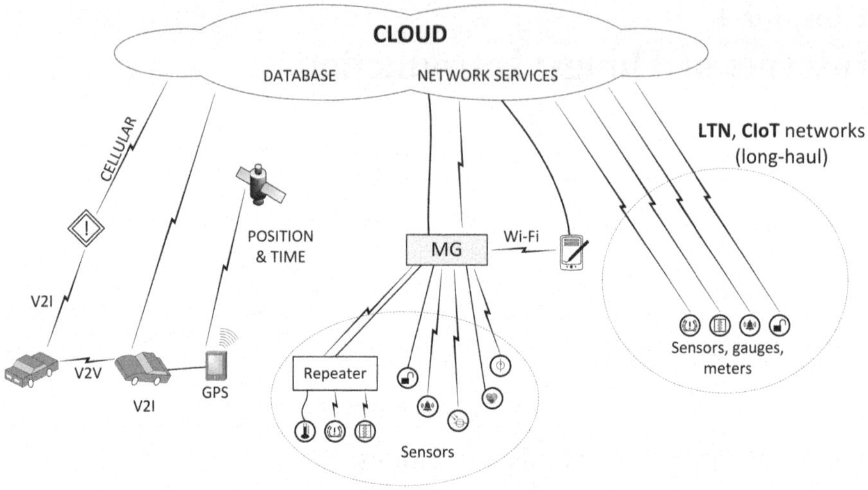

Fig. 1.1 A schematic concept of the Internet of Things in coexistence with current ICT systems

The idea of IoT is presented in Fig. 1.1, allowing at the same time to see the coarse division of the Internet of Things systems into long-range and short-range segments. Among the far-range segments, there are CIoT cellular and other solutions known as LTN (Low-Throughput Networks), which will be discussed, respectively, in Sect. 2.3 and Chap. 8 further in the book. It is the cloud in which both network and database services are operated that provides a shared space for data inflow, whether from traditional systems (cellular, wideband Internet access, etc.) or from systems in which information is generated by sensors, gauges, counters and meters. The ushering into the Internet of Things era brought about modifications in this area, including the emergence of the concept of 'fog computing', more effective from the energy and transmission point of view than computing in the cloud, as well as the 'edge computing', 'ambient intelligence' and others, as will be described in Sect. 1.5.2.

Early systems enabling M2M-type communication, also called massive machine-type communication (mMTC), were designed to provide communication at short ranges: from a few meters to several dozen meters. The first empowering solution was Bluetooth [9], especially in version 1.2, successfully used until today (for connections between computer peripheral devices or headsets and loudspeakers); the IEEE 802.15.4 standard, commonly called (though not really correctly) ZigBee [10, 11]; or IEEE 802.15.6, used for creating connections in the immediate vicinity of or inside the human body (the so-called WBAN systems).

From the service and technological point of view, the Internet of Things structure consists of the so-called vertical and horizontal domains. The vertical domains can be identified with specific application areas and related companies and standardization processes (although the scale may differ) taking place within each of these areas. Horizontal domains, in turn, are specific for telecommunication systems used to implement assumptions and tasks defined in the vertical domains, creating real communication solutions and thus constituting a shared foundation for both categories.

In [12], seven basic vertical domains have been specified, constituting the Internet of Things application core and the main drive of its development:

1. **Smart cities**: contemporary cities should evolve towards interconnected ecosystems in which all components (energy, mobility, buildings, water management, lighting, waste management, environment, etc.) interact for the benefit of their inhabitants. This process is possible thanks to the use of IoT systems, taking priority for security and privacy and reducing negative environmental impacts in a reliable and scalable manner.
2. **Intelligent environment**: this domain covers all aspects of the immediate human environment which support monitoring health (with the emphasis on the elderly and the disabled), but also help people to stay professionally active and in touch with the community. The domain extends from health and social care to home and building automation.
3. **Intelligent agriculture and food production**: the application of IoT technology to all stages of the agricultural cycle is expected to promote its optimization and, at the same time, improve the quality of food. Automatic activities such as acquisition of large amounts of data, data processing, analysis and inference supported by the associated automation are to serve that purpose.
4. **Intelligent WBAN devices**: integration of intelligent systems with clothing, materials, patches, attachments, watches, etc. opens up a large field for new applications and benefits. The user is thus offered access to such technologies and services as nanoelectronics, organic electronics, sensors, actuators and positioning. The need to overcome a number of controversies in such areas as ergonomics, privacy, data security, reliability and health (e.g. in the context of protection against radiation occurring in the immediate vicinity of the body) should be taken into account here.
5. **Intelligent transport/automotive**: another broad area covering vehicle fleet tracking, use of self-propelled drives, monitoring vehicle operational parameters, intelligent transport systems which control traffic distribution in cities and traffic lights or availability of parking spaces, and advanced applications including automatic communication between vehicles (contributing to significant increase in transport safety levels).
6. **Intelligent environment (intelligent water resources management)**: the IoT will be a key component for solutions dedicated to vertical applications, such as environmental monitoring and control, using sensors supporting environment protection by way of monitoring air (smog detection), water quality/quantity, water supply infrastructure and its management, atmosphere and soil conditions.
7. **Intelligent industry**: in order to meet the ever-growing demand for industrial products, the market will need to implement mass-scale intelligence mechanisms based on the IoT solutions, creating a network of connected objects to facilitate measurement, monitoring, managing power/energy/raw materials and both wireless and wired communication (Fig. 1.2).

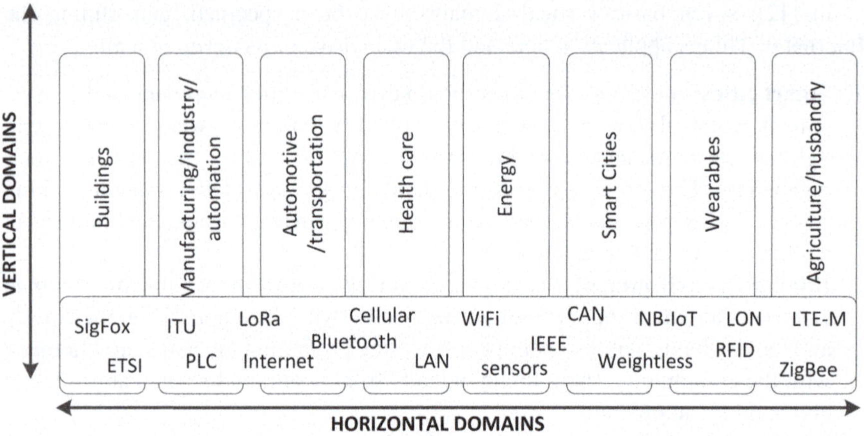

Fig. 1.2 Division of the Internet of Things into domains with an indication of standardization initiatives (based on [13])

Due to the type of application of Internet of Things technologies, it seems that from a utilitarian point of view, it can be divided into two main trends: mass and critical (Fig. 1.3). The division line here is the quality parameters of transmission required in each trend. Mass IoT, related to the traffic generated for the needs of intelligent buildings, logistics, tracking and fleet management, crop monitoring, breeding and, broadly speaking, intelligent media measurements, requires the following:

– Low costs and very low-power consumption, ensuring maintenance-free operation of the device over the years
– Small volumes of data
– The ability to handle a large number of devices counted even in tens of thousands per a base station

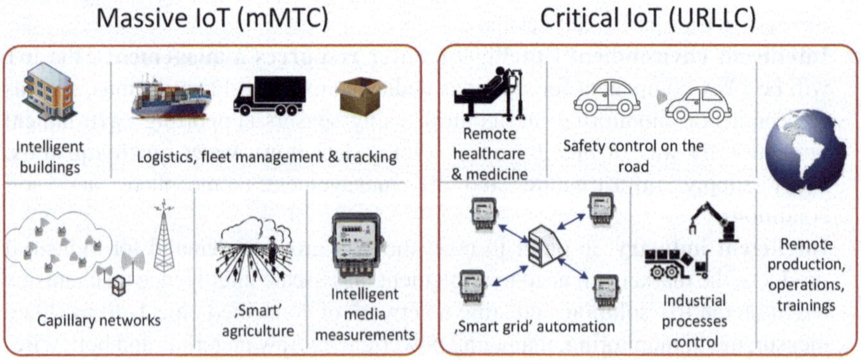

Fig. 1.3 A main classification of devices and the Internet of Things networks due to their service profiles

– High sensitivity (from 20 to 30 dB higher than standard wireless systems)

Critical IoT, however, is used for applications such as remote healthcare, interaction between vehicles on the road to prevent collisions, industrial processes control, remote production, etc. Although a large number of end users connect the critical IoT to mass IoT, the 'critical' variety differs considerably, mainly due to its requirements for transmission parameters, i.e.:

– Very high required reliability (low rate of undelivered frames), usually in excess of 'six nines', i.e. >99.9999%
– Very short delays, up to 1 ms, which is especially important in case of V2V (vehicle-to-vehicle) communication [14] or in medicine (e.g. during real-time operations)
– Relatively high throughputs being at least three orders of magnitude higher than in mass IoT, in which 1 kb/s is considered a typical reference value (see Table 2.3)

It should be emphasized that the division of the IoT systems presented here is part of a wider developmental trend determined by successive generations of cellular systems. For the upcoming 5th generation (or 5G), it is envisaged that systems falling into it will be used to support four main application groups aggregating 70 different utility scenarios initially defined in the 3GPP document [15]. Clearly, the first two of the following groups cover applications characteristic for the IoT systems intended for mass-type traffic (mMTC group) and critical traffic (URLLC group):

1. **Mass Communication** (the mMTC scenario – *massive MTC*) [16–17]: mass machine-type communication with a large number of devices (e.g. sensors, meters). The group is particularly suitable for new vertical services such as 'smart home', 'smart home appliances/peripherals/clothes', e-Health, etc. Recommended operating band: below 1 GHz, in order to easier satisfy the range requirements.
2. **Critical Communication** (the URLLC scenario – *ultra-reliable low-latency communications*) [18]: highly reliable low-latency communication in applications requiring low latency, high reliability, availability and security, e.g. in industrial control applications or Tactile Internet. It requires improved radio and haptic interfaces (see Sect. 1.5.1), optimized architecture, dedicated backbone and radio resources.
3. Enhanced **Mobile Broadband Communication** (the eMBB scenario) [19]: applications intended for fast data transmission, high user and base station density, high user mobility, support for traffic with highly dynamic transmission rates, convergence between stationary and mobile services and small cell size. Preferred work band: 24–100 GHz.
4. **Network Operations** (the NO scenario) [20]: refers to functional requirements such as flexible functionalities, creation of new values, migration and cooperation between networks (*internetworking*), optimizations, improvements and security.

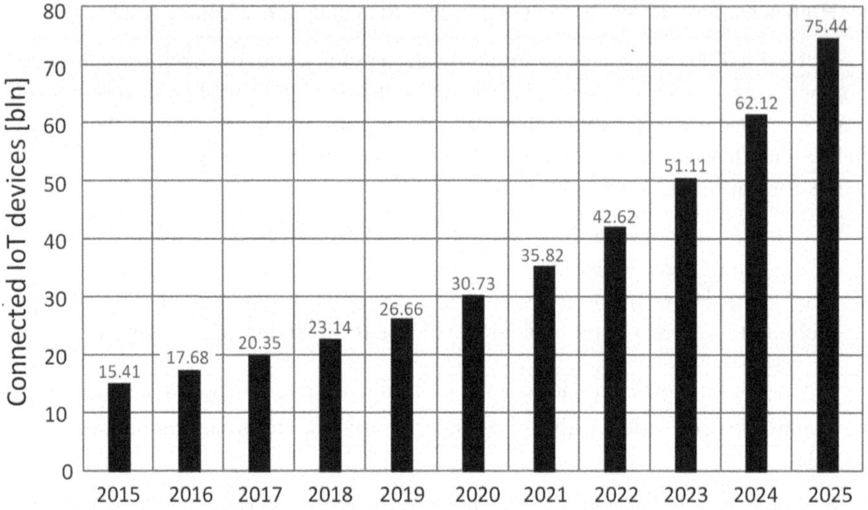

Fig. 1.4 Global trends: a growing number of connected IoT devices

As has been predicted in [21], the number of devices already connected to various networks, i.e. communicating with the environment, has reached about 20 billion. The exponential growth of devices reporting their readings in this manner or transmitting user data allows for an assumption that by 2020 this figure will approach 34 billion, while Cisco estimates indicate 50 billion [22] or even 200 billion (according to IDC and Intel [23]). KPMG, on the other hand, exercise more caution in their estimates [24], setting this number at approx. 28 billion devices in 2020. Following Statista.com [25], over 75 bln connected IoT devices are expected in 2025, to be reached with an exponential trend, as illustrated in Fig. 1.4, presenting the situation based on past data and actual premises.

1.2 Internet of Things: Definitions and Systematics

This chapter is based mainly on ITU Y.2060 recommendation [26], specifying the main functional and general architecture assumptions of the Internet of Things network as a phenomenon within a more general concept of *New-Generation Networks*.

1.2.1 The Internet of Things (IoT) Concept

According to the ITU definition, the Internet of Things should be perceived as a vision embracing far-reaching technological and social implications. From a standardization perspective, the IoT can be seen as a global information society infra-

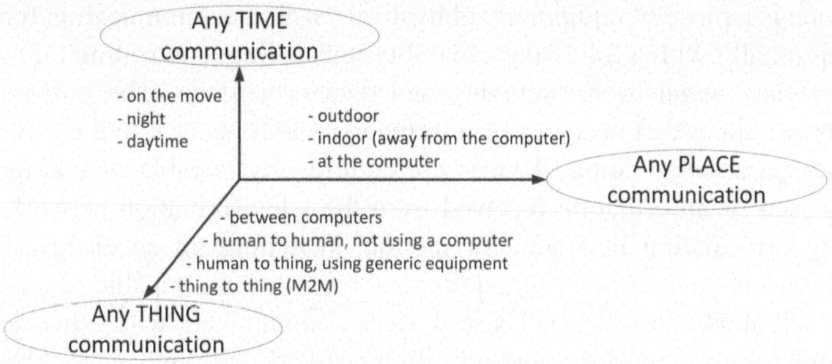

Fig. 1.5 The ICT domain dimensions taking into account IoT systems

structure, providing advanced services by way of combining (physical and/or virtual) 'things' with the help of existing and/or emerging *Information and Communication Technologies* (ICT). Using the identification, data acquisition, data processing and communication functions, the IoT enables full employment of 'things' to provide services for various applications, while ensuring security and adequate privacy. Thus, the IoT aimed at causing integration of leading technologies, including those related to communication between M2M (*machine-to-machine*), autonomous networks, data mining and making autonomous decisions, cyber security and privacy protection, cloud and fog processing and advanced sensor and actuator systems (such as actuators or controllers). These features therefore merge the existing functional space of ICT systems, i.e. communication 'at any time' and 'anywhere,' with another dimension, i.e. communication of 'any things', as illustrated in Fig. 1.5.

1.2.2 Technical Description of IoT

'Things', in the context of the IoT, are physical world objects (physical things) or information technology (virtual world) objects which may be identified and integrated in telecommunications networks. 'Things' should therefore be linked to information, be it static or dynamic.

Physical things exist in the world of physical phenomena and can be measured, excited (activated, controlled) and communicated with. Examples of physical things are the surrounding environment, industrial robots, consumer goods or electrical equipment.

Virtual things are, in turn, defined within the IT world and can be stored, processed and shared. Examples of virtual things are multimedia content and application software.

A physical thing may have its representation in the IT world in the form of one or more virtual things, by means of projection, while a virtual thing may exist without reference to any physical representation.

A device is a piece of equipment, obligatorily with a communication functionality and optionally with a following capability: measuring, controlling (activating or initiating), data acquisition, archiving and data processing. The purpose of the devices is to collect various types of information and transfer it to the ICT network for further processing. Some devices are additionally capable of taking specific actions, based on information received from the teleinformation network, e.g. by translating information into actuator motion, switching on an electrical circuit, changing settings of a measuring equipment, etc. (devices capable of such operations are actuators/controllers). These devices communicate with other devices in the following ways: via telecommunication networks through gates (case 'a' in Fig. 1.6), without the use of gates (case 'b') or directly with omission of any telecommunication network (case 'c'). Naturally, there are various combinations of these three scenarios, for example, devices can communicate directly via a local network (i.e. a network providing communication between devices themselves or between devices and gateways, e.g. in ad hoc networks), as in case 'c', or communicate via telecommunication networks using local network gateways (case 'a'). It should be noted that although Fig. 1.6 shows only the interactions taking place in the physical world (i.e. between devices), these interactions may also take place in the IT world (information exchange between virtual things) and between the physical world and the IT world (information exchange between physical and virtual things).

There are a number of types of IoT applications, e.g. 'intelligent transportation systems', 'smart electricity measurements' (smart grid), 'e-health' or 'smart home' (these scenarios will be also discussed in Sect. 2.3.1). These applications can be built on proprietary (company owned) application platforms or on generally available service

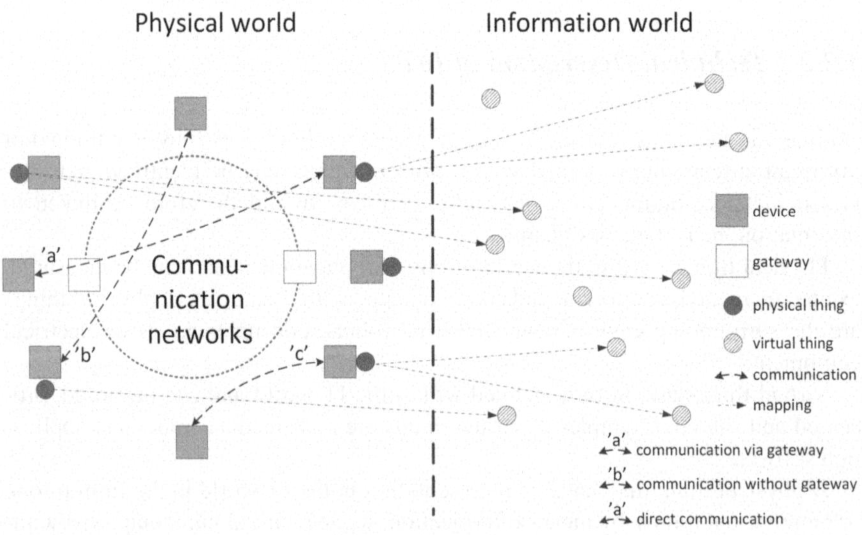

Fig. 1.6 Overview of objects and communication relationships in the IoT

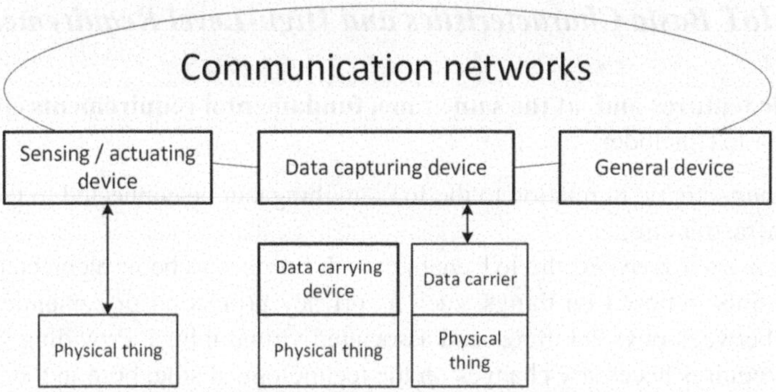

Fig. 1.7 Types of the IoT devices and their relationships with physical objects

and application platforms which provide generic services such as authentication, device management or billing. Telecommunication networks transfer data captured by devices to applications and to other devices, as well as instructions from applications to devices. Communication-related technologies, therefore, provide functionalities indispensable for reliable and efficient transfer of information. The IoT network infrastructure itself can be implemented either by means of existing network solutions, such as conventional TCP/IP networks, or by new network solutions, such as those described in later chapters or LTN/CIoT family systems. Figure 1.7 shows different types of devices and relationships between devices and physical objects. It is mandatory for all IoT devices to support communication functions with four main categories in this respect:

- *A data carrying device* attached to a physical thing, to ensure its indirect connection to a telecommunication network.
- *A data capturing device* with a read and write functionality (uploading), capable of interacting with physical things, while interaction can be indirect, i.e. via data transfer devices, or direct – via media attached to physical things. In the former case, a data collecting device reads the information from a data transfer device, optionally uploading information received from the telecommunication network to the data transfer device. There are several physical media employed in interaction between devices (both for transferring and downloading data), including radio waves, infrared, optical waves or galvanic connections.
- *A sensing and actuating device*, capable of detecting or measuring information about the surrounding environment and then converting it into digital form. Such a device will also translate electronic digital signals from teleinformation networks to specific operations, as is in the case of actuators. In general, local network devices communicate with each other using wired or wireless technologies, with or without the use of gateways to connect to networks.
- *A generic device* with built-in functionalities: processing and communication, capable of communicating with telecommunications networks using wired or wireless technologies. These devices include equipment and appliances typical for various domains of IoT application, such as industrial machinery, home appliances and smartphones.

1.2.3 IoT Basic Characteristics and High-Level Requirements

The basic features and, at the same time, fundamental requirements characteristic of the IoT include:

- *Interconnectivity*: in relation to the IoT, anything can be connected to the global ICT infrastructure.
- *Thing-related services*: the IoT enables such services to be implemented within constraints imposed on things, such as privacy protection or semantic consistency between physical things and associated virtual things. Providing such services requires necessary changes on the technological side, both in the physical and the IT world.
- *Heterogeneity*: the IoT devices are inherently heterogeneous, the reason being the diverse hardware platforms and networks they are designed to work with. Moreover, they are able to interact with other devices and service platforms via different networks.
- *Dynamic changes*: the state of devices can change dynamically, e.g. by going from sleep to awake or from a connection to breaking it. Variability may also occur in the hardware context, including location, speed or the number of devices.
- *Large scale*: the number of devices that should be managed and which require communication with other devices will be at least an order of magnitude greater than the devices currently connected to the Internet. Also, the ratio of communication traffic initiated by the devices as compared to the traffic generated by people will gradually increase (which is numerically expressed at the beginning of Sect. 1.1). Thus, the critical issue will be the management of generated data and their interpretation for the needs of the applications, which requires the use of efficient semantic mechanisms, but also data mining and processing techniques, wrapped up in a generic term, the Big Data.

High-level IoT requirements include:

- *Identification-based interconnectivity*: for IoT, the connection must include identifiers communicating devices. This also entails unified support for heterogeneous identifiers possessing, for instance, different syntaxes.
- *Autonomous network organization*: includes independent execution of the following control functions: management, configuration, repair, optimization and security, to be provisioned by the IoT networks. The purpose here is to adapt the network to various application platforms, communication environments and a great number of different types of devices.
- *Autonomous service provision*: the services should be capable of automatically capturing, transferring and processing data received from 'things', based on rules set up by operators, in some cases modified by subscribers. These services may be based on automated data fusion or data mining techniques.

- *Location-based services*: some services may be based on items or users location data. Both the location itself and its updating should take place automatically, taking into account local legal regulations, including laws protecting privacy.
- *Security*: in IoT, every ‘thing’ is communicated, which generates security-related threats in such areas as confidentiality, authentication or coherence of both data and services. The most difficult aspect is the security requirements integrating different security policies and techniques relating to various devices and networks in IoT.
- *Privacy protection*: this requirement exists due to the fact that many ‘things’ have their owners and/or users. The data measured, taken from ‘things’, can transfer restricted information which needs to be protected. The protection is required both at the stage of transmission, aggregation, archiving, exploration and processing. Moreover, the enforcement of these security rules should not at the same time be an obstacle to the process of data source authentication.
- *High-quality services related to human body*: depending on legal conditions of a given country, these services are related to the collection, transmission and processing data related to static and dynamic behavioural features of the human body, with or without human intervention.
- *Plug and play*: the functionality enabling the current semantics generation or downloading semantics-based configurations for the seamless integration and cooperation of connected ‘things’ with applications, while ensuring responsiveness appropriate to the requirements specified in a given application.
- *Manageability*: a feature necessary to ensure a normal operation of devices as elements of telecommunication networks. While IoT applications usually work in the automatic mode, their operational parameters should be controlled by appropriate management processes.

1.2.4 The IoT Reference Model

The IoT systems reference model presented in Fig. 1.8 consists of four layers and management/security functions related to those layers:

- An application layer
- A service and application support layer
- A network layer
- A hardware layer

The application layer contains IoT applications.

The service and application support layer contains two basic functionalities:

- *General support functions*: including functionalities shared by various IoT applications, such as data processing or archiving which may be triggered by specific support functions, e.g. building further specific support functions

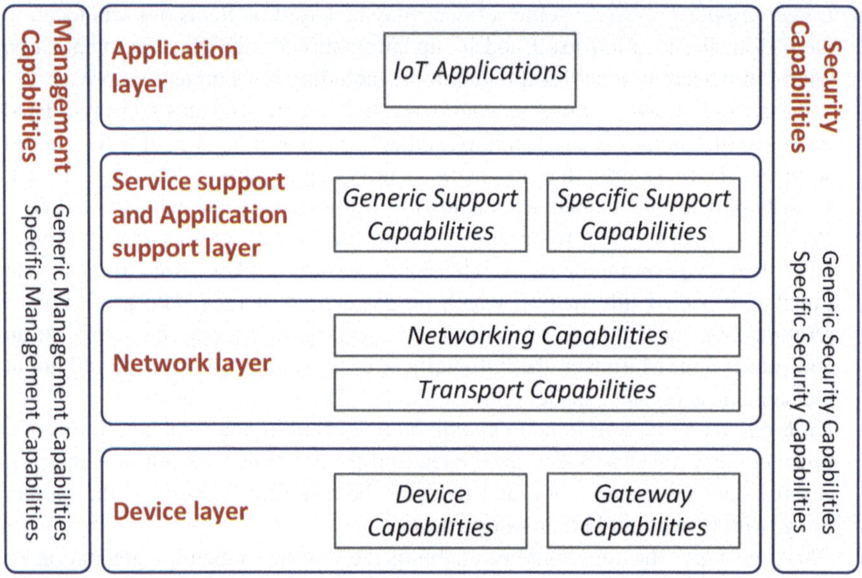

Fig. 1.8 The reference Internet of Things model

– *Specific support functions*: aimed at specifically defined IoT applications, possibly with an extension of further levels of specific support functions

The network layer consists of two types of functionalities:

– *Network functions*: providing adequate functionalities of network connections control, such as access to the medium or resource management, mobility management, authentication, authorization and billing (accounting)
– *Transport functions*: focusing on providing connectivity for information from services and applications transmission, as well as for transporting control and management messages related to the IoT

The hardware layer is logically divided into two categories of functionalities:

• *Device functions* – core functionalities, though do not exhaust all possibilities:
  – *Direct interaction* with the telecommunication network, i.e. the state in which the devices are able to collect data and send information to the network directly (i.e. without intermediate gates) and directly receive information (e.g. commands) from the network. The device may, but does not have to, simultaneously support the indirect interaction functions.
  – *Indirect interaction* with the telecommunication network, i.e. a condition in which the devices are able to collect data and transmit information to the network only via devices acting as network gateways and to receive information

(e.g. commands) from the network in the same way. A device may, but does not need to, support simultaneous functions of direct interaction.
  - *Ad hoc network*: devices can form an improvised network which facilitates network installation speed but requires them to support appropriate algorithms ensuring such scalability (including routing algorithms).
  - *Putting to sleep and waking up*: a functionality that significantly improves the energy efficiency of equipment operation. More on this topic, based on examples of cellular IoT systems, is presented in Sect. 8.2.

- *Gate functions* – include, although do not exhaust all options (the subject also discussed in Sect. 1.5.3):
  - *Multiple interfaces support*: within the device layer, the gate functionality requires support for devices connected to the gate by means of various types of wireless technologies, such as CAN (Controller Area Network) Bus, ZigBee, Bluetooth or Wi-Fi. At the network layer level, in turn, the gate functionality requires the ability to communicate with the access segment of such networks as PSTN (*public switched telephone network*), 2G-5G cellular generation, Ethernet or DSL (*digital subscriber line*).
  - *Protocol conversion*: two situations can be identified where the gateway functionality is required. One of these is the case when communication at the network layer level is carried out using various hardware protocols, e.g. ZigBee and Bluetooth. The second situation involves the case of various protocols being used within various network layers, for instance, when the ZigBee protocol is used in the hardware layer and the cellular protocol in the network layer.

Management functions. Just like in traditional telecommunication networks, the management functionality includes five main management items, fault, configuration, accounting, performance and security, sometimes referred to as FCAPS. Management functions are further divided into general and specific functionalities:

- General management functions include:
  - *Device management*, including remote activation, deactivation, diagnostics, updating of the firmware and system software and management of the device risk status
  - *Local network topology management*
  - *Traffic and load management*, e.g. by means of network overload detection or implementation of network resource reservation mechanisms for *time-critical* or *life-critical* transmissions

- *Specific management functions* are strictly connected to requirements of a specific type of application, e.g. for the needs of intelligent electricity metering services, i.e. smart-grid.

Security functions fall into two groups:

- *General security* functions, including:
  - At the *application* layer level: authorization, authentication, data confidentiality and protection of data consistency (integrity), privacy protection, security auditing and immunity to viruses
  - *At the network layer level*: authorization, authentication, confidentiality of user and signalling data and protection of signalling consistency
  - *At the hardware layer level*: authorization, authentication, device integrity validation, access control, data confidentiality and consistency protection
- *Specific security functions* are strictly related to the requirements of a specific type of application, e.g. for mobile payments.

1.3 IoT: Risks and Concerns

Based on the opinions gathered from representatives of various industries (not only heavy industry), whose needs form a main driving force behind the development of the Internet of Things, several key barriers can be identified which may discourage recipients from implementing these systems. These barriers, as were pointed out, for example, in a survey documented in [27], may be a potential obstacle hindering global adoption of IoT techniques and systems and include such factors as:

1. *Data security*: 40% of the surveyed users. The problem concerns the lack of certainty that the huge amounts of data, the acquisition of which is obviously linked with the IoT, will be safe, especially in the era already referred to in media, as the post-Snowden era. This is a significant problem, especially in the USA where, since 2015, the Federal Trade Commission may impose financial penalties on institutions in which, according to the Commission's judgement, the collected data is inadequately secured (including, e.g. hotels, banks, offices, etc.).
2. *Transmission security*: 40% of the surveyed users. The issue is related to the expected multiplicity of new systems in the field of IoT (long and short range), and thus it is difficult to monitor, and even more so to control and secure, the network of connections and routes of data flow.
3. *Significant implementation costs*: 38% of the surveyed users. Mainly associated with the necessity to adopt a business model based on cloud solutions to effectively use the benefits of the IoT systems. In this respect, certain hopes are brought by the popularization of the approach called Edge Intelligence, which largely reduces the amount of Internet of Things data actually transmitted via telecommunications networks (which will also be discussed in Sect. 1.5.2).
4. *Restricted knowledge about available solutions*: 38% of the surveyed users. The problem is neither new nor specific to the IoT and represents a rather frequent phenomenon accompanying early phases of implementing new genera-

tions of ICT systems. In short, users are often unaware of the potential and benefits of recent solutions, and, consequently, they will rather stick to the existing ones claiming that perhaps these are 'not the best, but at least well tested'. At the threshold of IoT reality, after several years of experience, a tendency can be observed in many telemetry companies to rely on 'old, good SMS/GPRS solutions' rather than invest in the Internet of Things techniques, in their view remaining a risky unknown. In the case of IoT networks, the problem may arise in the form of a significantly large number of systems developed almost simultaneously (LoRa, SigFox, Weightless-P, HC-12, NB-IoT, LTE-M (TC), etc.), which may simply be confusing for an average user. To a large extent, the standardization institutions and telecommunication regulators may contribute to breaking these cognitive barriers by conducting appropriate advertising campaigns, seminars, demonstrations, etc., thus encouraging companies to boldly switch over to the new IoT technologies.

5. *Lack of proper infrastructure*: 33% of the surveyed users. This is connected with still insufficient supply of IoT systems offered by service providers on the market. The providers often limit their services to specific geographic areas only, expecting to obtain capital from customers, which is meant to be an incentive for further development.
6. *Lack of standards*: 29% of the surveyed users. Some measures have already been taken in this area (of which, of course, a certain number of users may simply be unaware). This issue will be discussed in more detail in Sects. 2.1 and 2.3.
7. *Lack of interoperability between standards*: 28% of the surveyed users. A remedy in this case may be the use of integrating gates. Respective measures have been taken by parties involved in integrating solutions dedicated to, or adaptable to, performing calculations in the fog. One example is the OpenFog project founded by, among others, by Cisco, Dell, Intel, Microsoft and Princeton University, whose aim is to promote fog computing technology, assuming, as a starting point, that the requirements of the Edge Intelligence have been met. These requirements play a prominent role in the concept of the Internet of Things, which will be further discussed in Sects. 1.5.1 and 1.5.2.
8. *Uncertainty as to the real profits*: 27% of the surveyed users. Implementation of IoT systems also entails the need to implement various accompanying tools, e.g. data security and transmission mechanisms, management of an extensive database system archiving vast amounts of data, middleware software, application programming interfaces (API), Big Data, data analysis or data visualization methods. There is concern that the costs might significantly reduce the real income associated with the availability of such a large and detailed database provided by IoT systems. Many ICT companies are afraid of repeating the overrated revolution that RFID (radio frequency identification) systems were expected to initiate over a decade ago which led to the development of many new RFID applications areas, although their real market success was confirmed only in large-scale chain of store magazines (e.g. Walmart, Amazon) or shopping centres.

9. *Incompatibility of the current process and work flows with the IoT requirements*: 26% of the surveyed users. Concerns in this respect also stem from the general lack of experience of companies with Internet of Things systems, mainly in the scope of linking various data flowing through IoT channels from devices which, until now, were only to be read and did not generate any significant telecommunication traffic that would call for any special management and analysis methods. On the other hand, most corporations apply advanced methods of organizing work flow, logistics or production cycles, which they are reluctant to give up in favour of adopting solutions involving the IoT. There is a fear that – despite the awareness of the benefits of access to large amounts of data, previously less or completely inaccessible – this access may disrupt well-proven methods, introduce chaos or slow down decision-making processes resulting from the need to work on too much data. The number of available platforms (currently about 300, according to [28]) with potential to provide companies with comprehensive Internet of Things solutions seems too big, which adds to the confusion and disorientation of users as to the optimal choice of a particular option. Cleanly, this problem is basically a sum of all the doubts indicated above by the surveyed representatives of the business world.
10. *Immature technology*: 24% of the surveyed users. The argument that can partly be explained by insufficient awareness of existing IoT solutions, and which should partially be recognized as valid due to, for example, the lack of shared standards facilitating the adoption of a common methodology for implementing and maintaining all platforms. One cannot overlook the lack of reliable ICT platforms to audit IoT systems, both in terms of performance and cyber security.

New solutions in the field of ICT have always brought promise of measurable gains in virtually all branches of the economy. Their adoption, however, was inevitably followed by doubts, questions and threats. The situation is not different in the current phase (still regarded as initial) of implementing the Internet of Things solutions which are just one more step in the evolution of networks and ICT solutions. Among the abovementioned user concerns, several major ones, specific for IoT systems, can be identified:

- *Managing overwhelming amounts of new data* (not to mention security aspects) which is a challenge for both logistics and cyber security.
- *Insufficient knowledge about ICT solutions* (hardware and software platforms providing Internet of Things services).
- *Excessive fragmentation of solutions*, highlighted by the lack of standards which would enable universalization of approach or adoption of certain shared methods for implementing various IoT platforms, each of which currently seems to be unrelated to competing ones at any level, ranging from physical implementation of networks to data management models. Dedicated auditing platforms for IoT systems, expected to emerge in the near future, therefore arouse high hopes also in this respect.

1.4 New Concepts and Development Forecasts

Some solutions for the shortcomings discussed in Sect. 1.2 come with the development of concepts associated with the Internet of Things, which have developed in the meanwhile and which can be traced back to sensor networks, in particular the wireless sensor network (WSN). The WSN networks, whose rapid development was witnessed in the first decade of the present century, were the first solution to tackle the problem of transmitting very specific data. Two factors influenced the development of these systems: (1) the need for monitoring various parameters returned by counters, meters and sensors employed (mainly) in the industry and (2) technological maturity of telecommunication techniques and standards and their fairly high supply in the market such as the previously mentioned Bluetooth or ZigBee but also a considerable number of non-standardized niche solutions available at moderate prices. The most important achievement, currently implemented by the Internet of Things (although derived largely from short-range systems dominating in the WSN), is the perception of the need for limiting the amount of data transferred to the necessary minimum, preferably by delineating 'dense MTC traffic' areas and allowing only a small fragment of this entire traffic to leave these boundaries, containing only what is necessary and/or representative for the monitored phenomena, for example, results of statistics from measurement or summary reports, averages, confirmations, etc. This observation has given rise to such concepts as: ambient intelligence, edge intelligence and media gateways. There are direct implications of these concepts, embedded in the context and dedicated to the Internet of Things, namely, capillary networks and fog computing. These concepts will be the subject of a detailed discussion in Sect. 1.5.

The share of telecommunication traffic bound with the Internet of Things communication, as of year 2020, is presented in detail in Fig. 1.9, after Business Insider [21]. It is quite clear there that MTC communication has already gained a dominant position, although not in terms of the amount of information sent but rather in terms of the number of devices, with a forecasted 70% share in 2020.

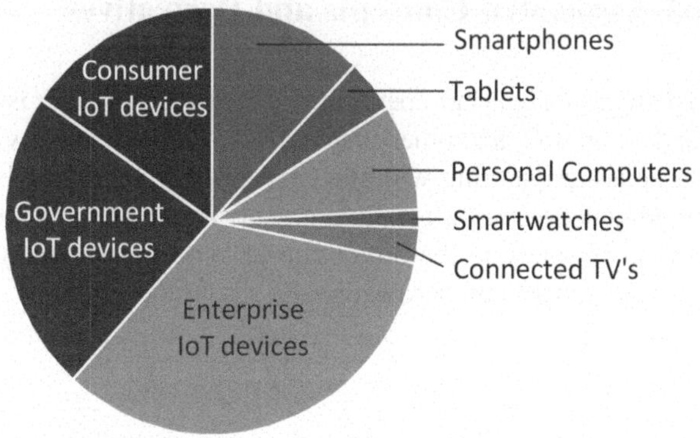

Fig. 1.9 Internet of Things devices current and forecasted market development (based on [21])

Due to the truly extensive IoT application market, it is worth quoting after [29] an anticipated share of 10 main segments of applications on the basis of global, publicly available information about planned or implemented IoT projects. Their number in this compilation though is greater than the seven key vertical domains listed in Sect. 1.1. The difference is only due to the fragmentation of some domains into subcomponents. For example, 'communications in industry' and 'other industrial applications' have been isolated from the 'intelligent industry domain':

1. Industrial connectivity, mainly in the mining industry: 22%
2. 'Smart cities' type projects: 20%
3. Smart measurements of media networks (smart energy/grid): 13%
4. Automotive (remote diagnostics, control and monitoring of the fleet, etc.): 13%
5. Other industrial applications: 8%
6. Agriculture (animal husbandry or crops monitoring): 6%
7. Intelligent buildings: 5%
8. Healthcare: 5%
9. Commerce: 4%
10. Chains of supply: 4%

The most intense activity in the area of creating new IoT projects is demonstrated by companies operating in the Americas, with a total of 44% share. Interest on the European side oscillates around 34% of the share, while the rest falls mainly to the Asia-Pacific region (i.e. APAC) and, to a lesser extent, the Middle Eastern and African countries (the so-called MEA region). The relationship between the geographic region and specific application segments also seems to be clear here: in the Americas, healthcare initiatives (61% of all IoT projects globally) and trade (52%) represent an extremely strong share of interest, while the majority of projects related to 'smart cities' is located in Europe (47%). The APAC region, however, focuses on investments in smart energy projects (25%) [29].

1.5 The IoT Associated Concepts and Derivatives

The Internet of Things is a concept operating in the context of other issues; some of them are founded on the IoT, and others derive from it. This chapter will be devoted to discussing these issues, starting with the concept of ambient intelligence (AmI) through edge intelligence to fog calculations already mentioned in Sect. 1.4. Among them, only AmI exists solely in the dimension of conceptual paradigm; the other concepts have specific impact on hardware and/or software solutions.

1.5.1 The Ambient Intelligence (AmI)

The source concept of AmI refers to sensitive responsive electronic environment, responsive to the presence of people, and refers to a certain vision of future consumer electronics, telecommunication and computing techniques. It was proposed by Eli Zelkha and Palo Alto Ventures team as a plan for 2010–2020 [30–33]. In fact, AmI devices cooperate to support people in their daily activities, tasks and foster habits in a simple but natural and transparent manner, using information and specific 'intelligence' obtained as a result of these devices being interconnected in a communicated telecommunications network. With reduced dimensions of the devices, their number participating in monitoring our environment increases, over time they become embedded in the environment (hence *embedded systems*), and user interface is their only point of contact enabling user communication. The concept of AmI refers to concepts such as *pervasive computing*, profiling, contextual awareness or designing new technologies and systems in an anthropocentric way, which are required to be:

- Embedded: many devices integrated with the environment
- Context-aware: capable of recognizing the user and the user's situational context (including gestures, emotions, habits, preferences, etc.)
- Personalized: functionality tailored to the user's needs
- Adaptive: capable of self-modification in response to the characteristics of the user's needs
- Anticipatory: capable of predicting the user's needs without the user's conscious intrusion or command

Originally, the concept of AmI was related to the home environment in which sensors, located in different parts of the home, were to report to the household members on the status of its various elements or by forcing specific actions without their conscious participation (e.g. placing an order in the grocery store to replenish the stocks, according to the owners preferences, turning on the sprinklers, lighting, heating, blinds, etc.) [34]. Over time, along with the advancement of the idea of the Internet of Things, the concept went beyond the limits of home environment to collect, analyse and report data on the wider environment (large-scale media readings, reporting on water quality, air quality, etc.). Effective operation of AmI therefore requires simultaneous presence and collaboration (according to the diagram shown in Fig. 1.10) of the following AmI actions [35], (1) sensing, (2) reasoning and (3) acting, including all necessary accompanying elements, (1) security and (2) computer interaction, with the user via a shared human-computer interaction (HCI) interface.

The monitoring consists in receiving information from the environment via perceptual devices in the form of sensors, detectors, meters, counters, etc., either

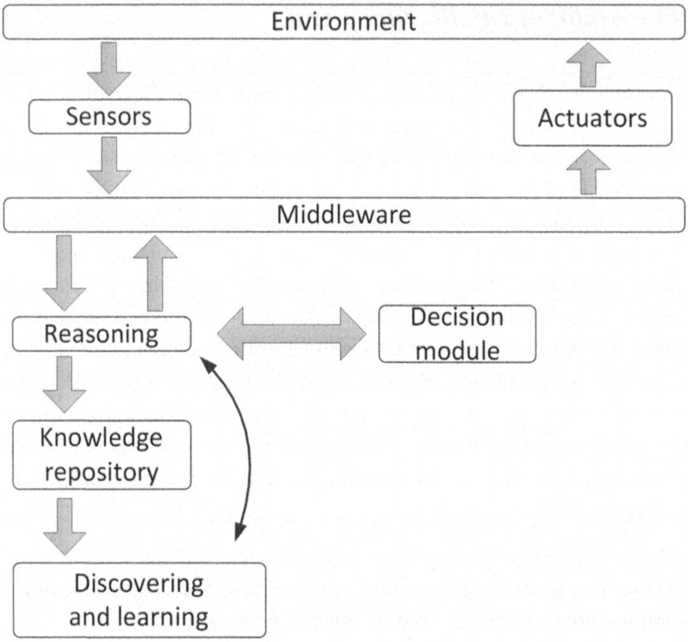

Fig. 1.10 Background intelligence machine (based on [35])

embedded in this environment (in the model case of AmI) or not. In the most general sense, sensors can be grouped with respect to their target applications: environment monitoring in open spaces (gas emissions, acoustic or electromagnetic smog, soil and water quality, etc.) and in closed spaces (air quality, temperature, humidity, noise, etc.); situational awareness, mainly to recognize a particular event context (e.g. motion detection, gesture recognition, velocity vector, acceleration, lighting detection, etc.); and monitoring structural and operational condition of buildings, including their mechanical and electrical parameters in order to increase the level of occupational safety within the building or in the course of constructing the building.

One of the most important challenges, mainly using embedded or battery-powered sensors, is the management of energy and memory resources, especially in the case of built-in or battery-operated sensors. There are two opposing concepts of data processing in this field: the centralized and distributed approach. The former requires transmission, via telemetric systems (consisting of a sensor and a transmission module), of all downloaded data to a common processing point, without any initial processing or analysis. The latter concept assumes initial in situ data processing, e.g. for the purpose of determining whether it is worth (in terms of the energy consumption) sending it; if the readings do not indicate deviations from user-defined thresholds, transmission may be discontinued to save energy and to reduce the amount of data generated. The concept of distributed computing, on the other hand, is closely related to the concept of 'edge intelligence' discussed later in this chapter.

Taking action is the result of a combination of obtaining reliable and comprehensive information on the studied phenomenon (acquired during the monitoring phase) and decision algorithms proposing actions to be taken based on this data, which may lead to optimization, or adjustment, or stabilization of the monitored/controlled phenomena. These actions can take place basically in two ways. The first is the direct interference of the automated system in the monitored environment, using actuators, for example, robots or controllers with the help of which it is possible to exert real influence on the parameters of the examined environment.

Monitoring and actuating are factors which mediate between intelligent algorithms and the physical environment. In order for these algorithms to be responsive, adaptive and thus work for the benefit of the user, it is necessary to add the **reasoning** functionality that can take a number of forms:

- *User profiling*, understood as the capability of modelling user behaviour, based on the type of measured data, the type of embedded model and the nature of the profiling algorithm.
- *Activity prediction and recognition*, i.e. the capability of predicting and recognizing activities occurring in the AmI environment, e.g. prediction of location and actions taken by household residents enables the AmI system to anticipate their needs and support them in their actions.
- *Decision-making* is the most evident and tangible manifestation of ambient intelligence presence because decisions are then passed to controllers (actuators), understood here as appropriate measures provided for in a given application to translate decisions into actions. An example of the above is an AmI system supporting a patient's health which, in the absence of actions on the patient's part, in response to deteriorating results returned by wearable sensors, are able to independently activate a telecommunication system (e.g. by making an automatic phone call to the hospital/call an ambulance) in order to report a probable loss of consciousness or any other condition possibly staving the patient off from taking rational decisions.
- *Space-time inferring*, related to user profiling, enriched with information tying the user's regular activity with specific locations in which these activities take place. Their occurrence in other locations or their lack in the locations in which they usually occur may be a reason for AmI to take certain decisions as to possible action scenarios. Such anomalies may indicate a potentially dangerous or undesirable situation for a monitored person.

Computer interaction with a user via the **HCI interface** should, first of all, make the user's daily life or professional activities easier. For this to become possible, HCI should be characterized by two key features:

- *Context awareness*, which is the starting point for building an AmI application. If the devices included in the ambient intelligence ecosystem are able to apply the latest technologies to conclude about the nature of the user's current activity (e.g. whether the user is walking or driving a car or is at home or at work) and his current surroundings, they will be able to intelligently manage both the content information and means of distribution of the information.

- *A natural interface* (point of contact). In the AmI environment, traditional methods of human communication with the machine (using a keyboard, touch screen, etc.) should be replaced with interfaces requiring a more subtle information transfer, all the same requiring AmI devices to be much more sophisticated in tracking motion, recognizing gestures and facial expressions or emotions and processing speech (and even murmurs, wheezing or whistling) to enable faultless interaction between natural human behaviour and intelligent environment. It implies that in order to facilitate the devices to correctly interpret a user's needs, the person cannot be in any way coerced to deviate from his/her natural behaviour, yet be certain that the message resulting from observing their activity will allow the AmI system to correctly determine the current context of the message. These types of contacts between the world of devices and the human world fit into the context of the so-called Tactile Internet, defined in ITU [36] and IEEE [37] as a network within the Internet, combining ultralow latencies (up to 1 ms) with extremely high availability, reliability and security. The point of contact ensuring such strict transmission parameters should, as originally intended, give the human operator the impression of direct, almost physical interaction with the operated machine, in accordance with the assumptions defined for the URLLC scenario in IMT-2020 systems (see Sect. 1.1). Intermediary technologies, including the visualizations of such interactions, are *the virtual reality* (VR) and *the augmented reality* (AR). In the Tactile Internet, translation of the human operator's actions into the reaction of the remote device, haptic interfaces are used, displaying the actuating element image (e.g. switch, lever, scalpel, etc.) and then translating the operator's selected action into the response of the real remote device thanks to fast transmission parameters necessary to perform the desired action, in the time close to real human perceptual abilities. The time is, of course, dependent on the sense involved in human-machine communication and differs by a sense involved in this interaction (acc. to ITU [36]): ~1 ms for the sense of touch, ~10 ms for the sense of sight, ~100 ms for the sense of hearing and ~1000 ms for the others monitoring-control interactions.

Adapting a direct human environment to their needs or preferences brings along many undoubted advantages. Generally speaking, AmI makes the user's environment more responsive and susceptible to his intended or planned actions, which reduces the physical effort necessary to perform those actions and thus offers more control over that environment. However, this control can be lost if the ambient intelligence system takes a wrong decision. Moreover, AmI technologies present a lot of potential security- and privacy-related threats. Security problems at the sensor level are associated, for example, with reliability of readings (possible defects or sensor incorrect location), error in processing, handling exceptions, anomaly detection, etc. In addition, if one assumes the wireless data exchange within the AmI system, there are realistic threats known also from conventional applications of radiocommunication systems, such as excessive signal attenuation, electromagnetic interference (intentional and accidental), possibility of eavesdropping (data theft), substitution of false information, etc. In the light of the above, there is always an inseparable issue of privacy, because correct profiling of a person, including their full behavioural context, requires gathering substantial amounts of information (stored not in one but in multiple locations, for safety reasons), successively supplemented with new incoming data facilitating the profile construction (or verifying the validity of the current one) which may undergo changes over time.

1.5.2 Edge Intelligence and Cloud/Fog Calculations

Cloud computing is a method of performing calculations based on an online database of computational resources. It is a model allowing ubiquitous, on-demand access to a shared pool of configurable computing resources (e.g. computer networks, servers, databases, applications and services) that can be delivered or released immediately with minimal effort [38]. Cloud computing enables individual and institutional users access to differentiated options of storage, processing or securing of their data in centres owned by entities providing such services, located away from the user's or institution's location (often in another city or country). The concept, therefore, is based on sharing resources to ensure consistency and economy of scale. Cloud computing is characterized by a wide range of services provided to users, such as:

- *Pay-As-You-Go* (PAYG), i.e. charging a user and scaling the payment according to resources actually used.
- *Software as a Service* (SaaS): a software licensing model based on subscription purchase for temporary use on a remote server, relieving the subscriber from the burden of having to purchase a full license for a higher price.
- *Education as a Service* (EaaS): making teaching resources available in the form of presentations, recordings of lectures, simulation software for learning, etc., for the purpose of organizing training or school classes.
- *Platform as a Service* (PaaS): providing customers with a remote system platform which allows them to run their own application without having to purchase their own system platform and update or make it secure.
- *Infrastructure as a Service* (IaaS): the provider delivers a full system base but also hardware to clients (servers, hosts, uninterrupted power supply, etc.), relieving the subscriber from the burden of maintenance or servicing.

Clouds can be divided into the following types from the standpoint of their organization:

- *Private clouds*, created and managed by a specific institution for their own needs, without making them available to other institutions as a service. Because it is the organization's own property, its maintenance is connected with incurring all costs related to this type of investment, including insurance, maintenance, updating or servicing, which makes this type of cloud relatively expensive in comparison with other types, with obvious benefits, however, on the data security side, resulting from better access control.

- *Public clouds*, located in the Internet, available to everyone for a fee, although often also on preferential terms (e.g. free access to archiving capacities of 10 or 20 GB), performing the full range of previously mentioned services (SaaS, PaaS, IaaS) in the PAYG model. High availability and ease of use of this type of cloud, however, come at a cost of a lower data security level, both at the stage of data transmission to the cloud and during the archiving phase.
- *Hybrid clouds* that involve the use of both forms in parallel but not simultaneously, switching between one and the other depending on the time of day, relying on the private cloud during top network traffic hours and reverting to the public cloud when the traffic slows down. This form of cloud also proves to be a good solution in foreseeable blackouts due to hurricanes, maintenance repairs of network installations, etc.

Although the services listed above are the best recommendation for the legitimacy of calculating in the cloud, this solution has some disadvantages which can include access to cloud resources depending on the Internet availability; data and privacy protection that depends on the strength of remote security, distributed systems and devices (implying that the end user is not in control of this aspect); and limited control and flexibility in the sense of limited user control over the functional infrastructure of the cloud.

Unprecedentedly large number of new devices which will be able to communicate in the network in the near future, as mentioned in Sect. 1.1, will require a huge capacity of telecommunication connections that carry all this traffic towards the cloud. Many of these devices will use (or already do) industrial protocols, other than the Internet Protocol (IP) to connect with controllers. Sending data to the cloud will, therefore, require translation into the IP. Moreover, many of these devices will send data in a continuous or periodic mode, which requires a quick analysis. One example may be temperature in an ironworks ladle, approaching the acceptable threshold. In this example, the actuator or driver correction has to be made in a short time. On the other hand, data transmission to the edge of the cloud for analysis may cause unacceptable delays in the response and eventually lead to the destruction of the pig iron. An obvious solution is, therefore, to place the functionality for data processing, computing and reasoning, as close to the point where this data originated as possible. This concept is called edge intelligence and consists in moving the computing cloud to the edge of the network. 'The edge' is defined as an interface where two of the networks' parts meet, namely, the transmission-computational layer and the sensor layer [39]. In practice, it requires separation of modules for concentrating data from a given area (referred to below as 'fog nodes'), performing necessary operations on the data according to predefined algorithms and taking decisions based on preprogrammed (policy) decision algorithms. Generally, there are two types of decisions:

- Sending a collective message (or messages) to the cloud, after it has been translated into a network protocol (e.g. IP), however in a much smaller number than the number of messages previously received from the sensors utilizing industrial protocols in the internal network.

- Triggering immediate reaction in the monitored environment by means of actuators, i.e. mechanical or electrical elements which control a given environment, capable of processing decisions within the edge intelligence decision-making system. Some examples of such a triggering are sending a signal to start cooling, sounding the alarm, etc. The signal is then translated into a physical stimulus initiating an action by means of actuators.

Support for small volume packets in IoT traffic and their different delay and transmission speed requirements have called for a new computational model, alternative to the cloud, taking into account the following new guidelines:

- Delay minimization: in case of applications in the production cycle, even millisecond delays could be important. The production loop often requires immediate feedback reactions for effective process control, which can only be ensured by placing a computational/decision module near the locations where measurements are made. One can say that minimizing of transmission delay is the most important argument in favour of the Edge Intelligence. In support of this argument, estimates can be quoted after [40] that by 2019, more than 40% of data generated by IoT devices will be stored, processed and analysed for the purpose of taking specific actions, near or on the edge of the network (i.e. on the border between the sensor network and data transmission network).
- Saving network capacity: extreme cases of environments monitored by sensor networks already send data in quantities that make it almost impossible to transmit this data to the cloud for archiving and processing by means of standard networks. For example, an offshore drilling platform generates 500 GB of data per week, while commercial jets generate 10 TB per every 40 min of flight. [41]. To a large extent, the data is important or even critical for efficient operation of systems they support, which makes their transfer to the cloud both impractical (due to quantity) and risky (due to possible delays and losses).
- Data safety: due to the importance of sensor data used to control systems, often sensitive and potentially dangerous to people and the environment, securing data both during their transmission and during archiving.
- Data reliability: IoT data will increasingly convey significant information important to human safety and critical infrastructure. Their consistency and accessibility, therefore, cannot be questioned.

Traditional cloud computing architectures do not meet all of the abovementioned requirements because sending all data from the network edge to the data centre for processing causes evident delays. Telecommunication traffic generated by thousands of IoT devices, on the other hand, may quickly exceed the transmission capacity of intermediate network devices. Looking from another viewpoint, industrial regulations and legal requirements regarding privacy protection may prevent the collection and storage of certain types of data (e.g. medical). In addition, cloud servers only communicate using the IP which makes them incapable of supporting the multiple IoT network protocols. The solution to these challenges is to analyse and take action based on the data close to where they are created. This concept, obviously connected to the previously mentioned edge intelligence, is called fog computing [41], a term introduced to the ICT nomenclature by Cisco.

The fog in this sense consists, therefore, of processing devices called fog nodes, whose locations cover all possible locations with network access, such as floor, ceiling or wall of the factory, electric line pillar, traction masts along the railway lines, a vehicle or an oil rig. Basically, any computing device capable of storing data and communicating via a network can be a fog node. These include, for example, industrial controllers, switches, routers, embedded servers or surveillance cameras.

As concerns the principle of operation, the best candidate for a fog node is the element used for aggregating traffic from devices that lies in their closes vicinity and that will later forward the accumulated data to other sites for analysis, as in Table 1.1:

- The data most intolerant to delays, e.g. information necessary to verify control loops or proper protection in the *smart grid*, are analysed in the fog node located closest to this data origin. Thus, any sign of a problem can be met with an immediate response of the fog node in the form of an appropriate command sent to actuators or controllers.
- Data which can wait seconds or minutes for action to be taken is sent to aggregation nodes for analysis. Invoking to the smart grid example, each substation could have its own aggregate node reporting the operational status of each power source located in the further part of the system.
- Data least sensitive to delays can be sent to the cloud for the needs of historical analysis, *Big Data* analyses or long-term archiving. For example, thousands or hundreds of thousands of fog nodes could send periodic summaries of data collected from the *smart grid* to the cloud in order to store or build long-term models.

To sum up and compare actions in the fog and in the cloud, the above discussion can be comprised in the following conclusions, systematized also in Table 1.2 and Fig. 1.11:

Table 1.1 Optimal IoT data analysis location

	The node nearest to IoT devices	Aggregation node	Cloud
Response time	Milliseconds up to <1 s	Seconds, minutes	Minutes, days, weeks
Examples of applications	M2M communication for the needs of haptics[a], including telemedicine and training	Visualization Simple statistics	Big Data analysis Graphic desktop
Data storage time	Temporary	Short term: hours, days, weeks	Months, years
Geographical area	Strictly local, e.g. within a single city block	Wider (cities, larger regions)	Global

[a]Haptic technologies employ mechanical communication with users through the sense of touch, using alternating forces, movements or vibrations

Table 1.2 Comparison of fog (boundary) calculations with cloud calculations [38]

Parameter	Cloud	Fog
Delays	High	Low
Jitter	High	Very low
Server nodes location	In the Internet	At local network edge
The distance between IoT devices and the server	Within a single hop or multiple hops	Within a single hop
Safety	Unspecified	Good, definable
Probability of an attack on data	High	Relatively low
Geographical distribution	Centralized	Decentralized
Mobility support	Limited	Supported
Interactions in real time	Supported	Very likely
Last mile connection type	Wired/leased line	Wireless

- Fog nodes:
  - Obtain data directly from IoT devices using one of many non-IP industrial protocols.
  - Support real-time control and analysis applications involving data obtained from IoT devices.
  - Provide short-term storage of IoT data, usually up to 1–2 h.
  - Process and filter data and build reports in order to reduce the amount of data transferred.
  - Send periodic data summaries to the cloud.

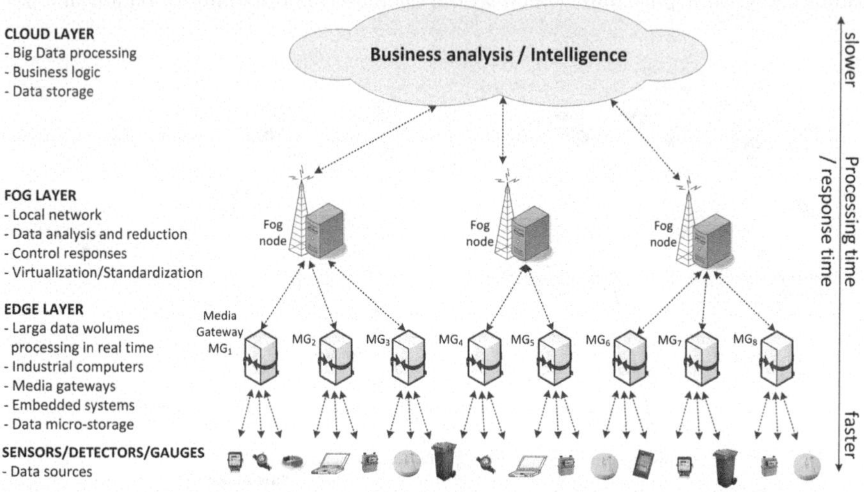

Fig. 1.11 Calculation layers in the cloud and fog model for the needs of IoT (based on: [35])

- Cloud layer or a cloud node:
  - Receives and aggregates data summaries collected from many fog nodes
  - Performs IoT data analysis and confronts outcomes with data obtained from other sources for the purpose of building business models
  - Has the capability of sending new guidelines to fog nodes regarding IoT data processing policy and decision mechanisms.

1.5.3 Capillary Networks and Media Gateways

The capillary network concept results from the already mentioned in Sect. 1.1 general division of IoT networks into long-distance and short-range architectural segments. The latter, like capillaries transporting blood to/from nerve endings, is called a capillary segment or network, as shown in Fig. 1.12.

In spite of the fact that one of the main goals set for IoT systems is to obtain the largest possible range, as compared to traditional (e.g. cellular) radio systems, there is a group of applications in which direct signal transmission from the place of its origin (e.g. from a gauge) is neither necessary nor recommendable. These applications include all kinds of measurements intended to draw an overall situational picture, in which only the final report based on a number of collected data is of practical value, instead of the information sent from all individual sensors, for example, the development of thermal maps for buildings which requires deployment of a large number of sensors in specific locations. In this situation, transmission of individual readings to a remote network server would require each sensor to have a separate radio module which, due to the presence of many similar signal sources, would be exposed to increased interference or collisions due to numerous simultaneous transmissions, which would decrease the reliability of data transmis-

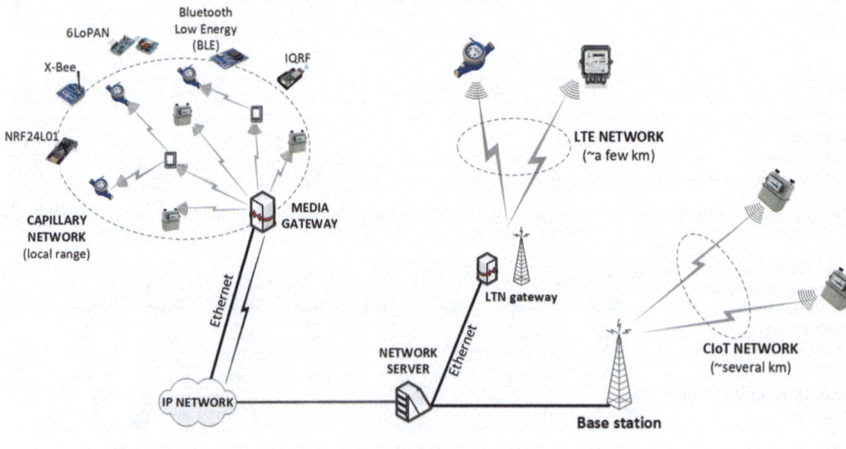

Fig. 1.12 Illustration of the capillary and long-distance network concept

sion. In addition, each reading would have to be independently registered in the database. Avoiding such situations was the motivating factor for the development of capillary network concept, in which relatively cheap and uncomplicated battery-powered devices communicate with a concentrating device, or a media gateway, in the star topology or in other configurations, including mesh connections. The market is currently abundant with solutions suitable for implementing this segment, offering numerous systems allowing transmission from several meters (e.g. Bluetooth BLE, FS1000A) to more complex arrangements and allowing transmissions in buildings at distances from a dozen to several dozen meters (e.g. IQRF [43], ZiGBee or NRF24L01).

Media gateways, or machine-to-machine (M2M) gateways, constitute a hardware implementation of the edge intelligence concept discussed in Sect. 1.5.2, which can generally be divided, with respect to the processing depth performed on the data received in the gateway from the capillary network, into the following categories:

- *Basic*, whose functionality is practically limited merely to translating protocols, i.e. from protocols natively adapted for the capillary network, e.g. ZigBee, Bluetooth, Z-Wave, IQRF, etc., to long-range protocols such as LTN, cellular or Internet protocols (e.g. TCP/IP).
- *Extended*, with a functionality similar to the assumptions defined for fog computing, i.e. capable of analysing and processing data collected from the capillary network, then their filtration according to significance of the monitored environment/phenomenon, securing and eventually translating short-range protocols (inbound in the capillary segment) into long-range network protocols (e.g. IP) or cellular networks (e.g. LTE or 5G).

In keeping with the assumptions of edge intelligence, the role of M2M gateways can therefore be gathered in the following list:

- Ensuring cooperation and integration of many different domains into a single platform (as in Fig. 1.13). The task primarily includes bidirectional conversion of protocols between the capillary and long-distance segments mentioned earlier.
- Aggregation of data and their local processing and reacting to events, due to which a faster response to events is possible, a higher probability of obtaining a set of correct (uninterrupted) data from all devices/sensors constituting a given telemetric network, with lower transmission and archiving costs.
- Securing data and networks, including data encryption, management of SSL/TSL certificates, authentication, VPN communication, firewalls and control of launched applications by performing so-called whitelisting.
- Providing reliable communication with IoT platforms based on cloud communication, by giving up traditional network protocols (e.g. XML, HTTP, TCP) in favour of the 'slimmed-down' communication protocols, dedicated to telemetric systems (or more generally – to the IoT) with low transmission speed, optimized in terms of overhead minimization, such as EXI (*Efficient XML Interchange*), CoAP (*Constrained Application Protocol*), MQTT (*MQ Telemetry Transport*),

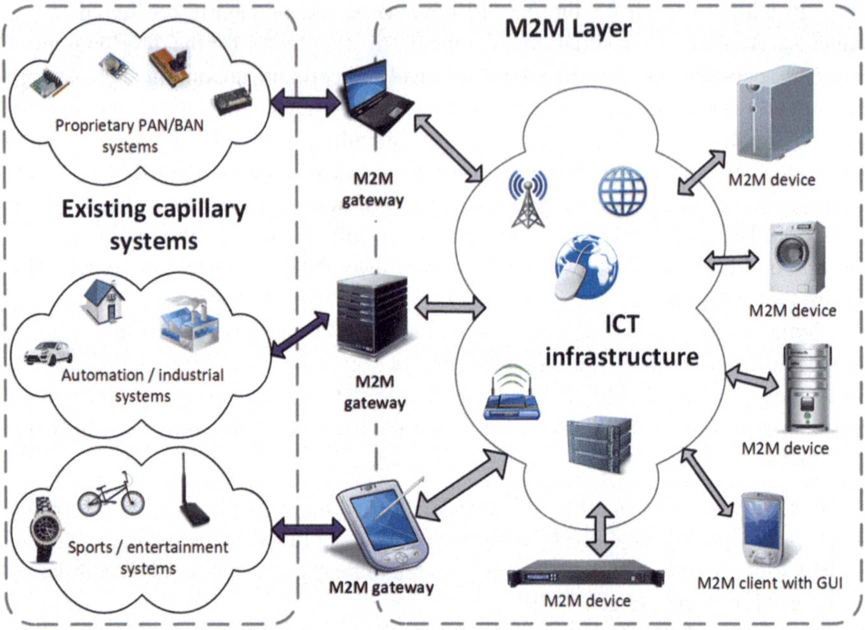

Fig. 1.13 Illustration of the media gateways role in mediating between sensor domains and the ICT infrastructure

6LoWPAN (*IPv6 over Low-Power Wireless Personal Area Networks*) or UDP (*User Datagram Protocol*). The measurable effect of such optimization is the reduction of packet sizes from hundreds/thousands of bytes (in traditional protocols) to tens of bytes (for the IoT protocols).

- Managing end devices in the capillary network segment and communicating with them by facilitating, e.g. remote firmware updates, scaling the number of devices in this segment (i.e. adding, subtracting or modifying/substituting) or intermediating in the two-way communication of these devices with the cloud.
- Local data archiving which has a bearing on savings on transmission costs, remote databases as well as increase in their protection, e.g. against unauthorized interception during transmission via public networks.

Long-range LPWAN/LTN or CIoT networks (e.g. LoRa, SigFox, Weightless-P, NB-IoT), on the other hand, provide dedicated solutions for meters/gauges associated with, for example, individual subscribers, whose readings are thoroughly representative of their user profiles based, for example, on pricing or optimization of the media delivery. Such data usually do not require aggregation or group processing, which justifies their direct transmission to the LTN gate (or base station). As a rule, LTN are therefore used for applications in the area of so-called smart metering, described in more detail in Sect. 2.1. Typically, these networks have built-in mechanisms allowing to 'recover' 20–30 dB in the link energetic budget for the purpose of compensating for additional losses resulting from penetrating the interiors of buildings, basements, etc. or to increase the outdoor coverage of traditional systems such as GPRS.

1.5.4 *A Statistical Traffic Model for the Machine-Type Communications*

The statistical model of telecommunications traffic is a useful tool in analysing the cost of technological capabilities of the user equipment (EU), related to the environment in which it is expected to operate. From an engineering point of view, correct traffic modelling is also necessary for the qualitative assessment of other aspects, such as the spectral efficiency or sufficiency of spectrum resources to handle all notifications.

According to 3GPP TR 36.888 (Annex A) [44], delays in packet delivery should not be higher than those in (E)GPRS and, if possible, comparable to LTE. Assuming the link budget larger by 20 dB than with traditional LTE users, an acceptable delay in response to the call has been set to 10 s, while in the case of emergency notification, the delivery time is 5 s. As a reference point for intelligent metering (see also Sect. 2.1), three traffic scenarios are assumed, presented in Fig. 1.14:

A. *Triggered reporting* between the base station and the EU subscriber device: ~ 20 bytes per command in the *downlink* (DL) and ~ 100 bytes for the response in the *uplink* (UL) with a 10-s Round Trip Time (RTT), i.e. between sending a command from the base station to the moment a response has been received.
B. *Emergency traffic* initiated by the UE module as a result of an extraordinary event – usually associated with exceeding the tested parameter, e.g. gas, water, electricity and certain threshold – reported using packets of size ~100 bytes, with a delay of 3–5 s from an event of the report generation to its reception by the base station. It is assumed that such events will occur sporadically, not more frequently than once a day. This kind of traffic has a risk of overloading the telemetric network, because in the event of an urban-scale technical breakdown (e.g. water supply or energy infrastructure failure), one can

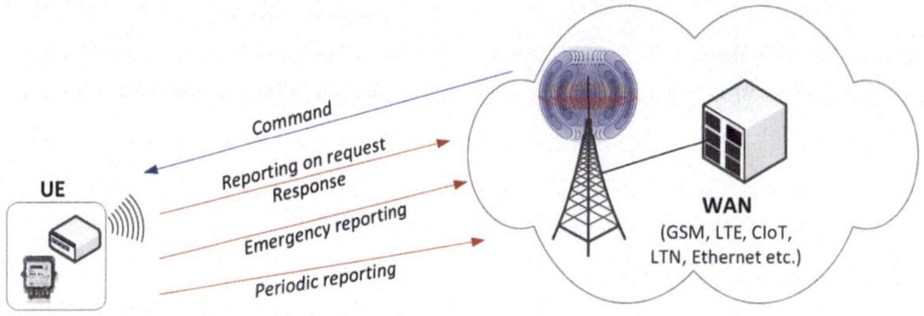

Fig. 1.14 Basic communication elements in networks carrying MTC traffic

expect a storm of reports from a large number of UEs reporting the same exceptional situation.

C. *Periodic reporting* or *keep-alive reporting* carried by messages of the size of ~100 bytes (UL), either insensitive or very tolerant to delays, e.g. with a 1-h tolerance. Typical for readings from various kinds of meters, e.g. gas, electricity, water, etc. A typical reporting frequency is once a day, week, month or less.

The model is based on investigations performed in London and Tokyo, specifying the household density per km^2, a recommended Inter-site Distance (ISD), and the resulting number of households per single cell. Currently, the original number of the UE per a single household has changed from the initial 3 UEs (acc. to [44]) to 40 UEs (acc. to [45]), which will be a basis for functional calculations in cellular IoT systems presented in Chap. 8. These assumptions were presented in Table 1.3, while the basic data for the traffic model itself were presented in Tables 1.4 and 1.5. It ought to be added that recent years have verified and helped modify traffic assumptions adopted in [44], which has resulted in emergence of the so-called mobile autonomous reporting (MAR) model, developed mainly for mobile IoT systems, presented in detail in Sect. 8.1 (Table 8.4), that serves as a valid reference for IoT systems planning.

Table 1.3 Auxiliary parameters for the Internet of Things traffic models [44]

Parameter	Highly urbanized area	Built-up area
Household density per km^2	London: 4275 Tokyo: 7916	London: 1517 Tokyo: 2316
ISD [m]	500 m	1732 m
Number of UEs per household	40 (acc. to [45])	
Number of houses per a cell	London: 926 Tokyo: 1714	London: 3941 Tokyo: 6017

Table 1.4 Traffic characteristics of periodic (regular) reporting of MTC devices [44]

Type of mobility	Frequency of UL reporting	Packet size	Type of mobility
Stationarity	1 m (optional), 5 m, 30 m, 1 h	1000 bytes, 10,000 bytes (optional)	Stationary, pedestrian (optional, no requirement to transfer traffic between base stations)
Limited mobility	5 s (optional), 10 s, 30 s	1000 bytes	Vehicles (no requirement to transfer traffic between base stations)

Table 1.5 An MTC statistical traffic model parameters [44]

Traffic model parameter (UL/DL)	Value
Report sizes (called out)	256 bytes, 1000 bytes
Distribution of interval between reports	Exponential (Poisson) with an average of 30 s

Chapter 2
IoT Networks Standardization and Legal Regulations

2.1 The Smart Metering Concept

The first real step towards conscious energy management was European Parliament Directive 2006/32/EC [46] on the energy end-use efficiency and energy services. This document, in the form of a recommendation, addressed the members of the European Union asking them to search for new sources of sustainable energy and introduce mechanisms for saving energy obtained from traditional, i.e. nonrenewable sources. Pessimistic forecasts as to the amount of raw materials available, with the ever-increasing demand for energy and the desire to reduce the emission of carbon dioxide to the atmosphere, were the motivation behind this appeal. Special emphasis was placed on the issue of energy efficiency; the phrase is often repeated in the directive, defined as 'the relation between real performance, services, material goods or energy and the energy put into their creation'. The directive encourages promotion of energy efficiency not only among large industrial plants but also small businesses and individual consumers often underestimated as energy users. A measurable goal was to achieve at least 9% savings on energy consumption owing to the following actions: energy audits, implementation of a nationwide monitoring system for energy consumption or educating consumers in the financial benefits of sensible utilization of energy by dynamic and adaptive tariff adjustment to real needs of individual or institutional users. By so doing, the foundations for today's prosumer market were laid, in which 'prosumer' means a user of electricity responsibly collaborating with the supplier with the view of rational energy management. Moreover, the mere combination of the words 'producer' and 'consumer' also implies an option of energy production by the consumer, e.g. in domestic power micro-plants (wind- or watermills), which contributes to the increase of the state energy budget.

The idea of 'smart metering', based on the above directive, along with the proposal for this term emerged in 2009 as part of a standardization mandate [47] issued by the European Commission to the following institutions: CEN *European*

K. Staniec, *Radio Interfaces in the Internet of Things Systems*,
https://doi.org/10.1007/978-3-030-44846-2_2

Committee for Standardization, CENELEC *European Committee for Electrotechnical Standardization* and ETSI *European Telecommunications Standards Institute*, in the area of measuring instruments for the development of open media architecture using communication protocols enabling their interoperability. A general goal of the mandate was to develop European standards for reading and transmitting data from meters (water, gas, electricity and heating) intended to raise the awareness among users of these media of the actual consumption and – on this basis – optimize the supply from the recipient's financial viewpoint (bills for the real instead of forecast or estimated consumption), as well as from the national energy management viewpoint. The issue of *smart metering* is described in detail in ETSI documents referring to particular areas in which monitoring and data measurement reporting systems are defined for:

- ETSI TS 102689 [48] – describes M2M services requirements, the aim being the provision of reliable and cost-effective communication for a variety of applications. The document consists of the following chapters: (1) general requirements and characteristics of M2M connections; (2) management and specifications for management models, including necessary functionalities such as failure detection, remote configuration, billing, etc.; (3) functional requirements stipulated for M2M services, including requirements for data acquisition and storage, reporting schemes, remote control operations, etc.; (4) security issues, including authentication of M2M devices, consistency and security of data, confidentiality, etc.; and (5) naming, numbering, device identification and addressing schemes specific to M2M devices.
- TS 102921 [53] – specifies main interfaces expressed in 'xIy' format, where 'x' and 'y' refer to 'a', application; 'd', devices; and 'm', core part of the M2M network. They specify protocols/APIs, frame format (data model) and their respective coding. These interfaces have mechanisms enabling cooperation between the two main components of the M2M architecture, i.e. the network domain and the devices/gateways domain, also in their perimeter, as shown in Fig. 2.1.
- ETSI TS 102690 [49] – describes the entire architecture (Fig. 2.1) of a generic smart metering system, specifying its individual functional blocks and reference points between them. This architecture assumes the use of the IP network provided either by wired or cellular service providers.

At the most general level, it is assumed that M2M structure is divided into two domains: (1) devices and gates and (2) network [50]. The former consists of:

- An **M2M device** communicating with the network domain via the M2M gateway or directly, if it has a built-in functionality for the provision of M2M services and applications. In the latter case, the metering module integrated with one of the long-range LTN/CIoT techniques is the solution enabling direct communication, as will be discussed in Chaps. 7 and 8.
- An **M2M gate** being a point of concentration for telemetric traffic which originates from one or more M2M devices connected via the local M2M network. It should be noted, however, that the functional scope of the M2M gateway (also

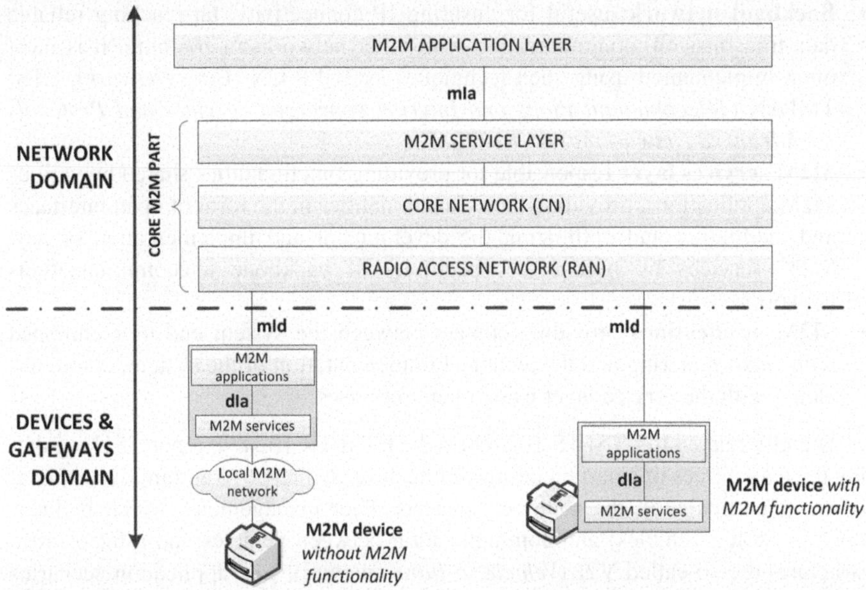

Fig. 2.1 High-level architecture of the M2M system (according to [49])

referred to as a *media gateway* in the context of the Internet of Things), as already stated in Sect. 1.5.3, is by no means limited to protocol translation and inter-domain 'portability' of M2M traffic. These gateways play a prominent role in the implementation of the edge intelligence concept whose requirements include traffic filtration, preprocessing, data protection, etc.

- **A local M2M network** enabling communication between M2M devices and M2M assigned gateway. In practice, this network is implemented, for example, in the form of a wireless sensor network (using standards such as Bluetooth, ZigBee, IQRF, Wireless M-Bus, NRF24L01, FS1000A, etc.) or in the form of wired networks, such as CAN, M-Bus, PLC or KNX.

The other, network domain, includes:

- The **access network**: an intermediary entity between the media gateway (M2M) and the backbone network. In this regard, cellular radio access systems, especially 2G and its derivatives (SMS, GPRS), are traditionally popular solutions due to relatively large ranges and area coverage higher than other generation systems, such as UMTS (3G) or LTE (4G). Typical access networks include xDSL (*Digital Subscriber Line*), HFC (*Hybrid Fibre Copper*), satellite access, GERAN (*GSM EDGE Radio Access Network* w GSM), UTRAN (*Universal Terrestrial RAN* in UMTS), eUTRAN (*evolved UTRAN* in LTE), WLAN (*Wireless Local Area Network*) or WiMAX (*Worldwide Interoperability for Microwave Access*).

- **Backhaul network**: useful for ensuring IP connectivity, far-reaching reliable data transmission, communication with other networks or roaming. It is most often implemented using such techniques as 3GPP CN (*Core Network*), ETSI TISPAN (*Telecommunications and Internet converged Services and Protocols for Advanced Networking*) or 3GPP2 CN.
- **M2M services layer** responsible for providing functionalities shared by various M2M applications, providing these functionalities in the form of open interfaces and facilitating and optimizing the development and implementation of new M2M services by hiding the specificity of backbone telecommunications networks.
- **M2M applications** providing contact between the system end user equipped with smart metering and the technical implementation of the system, communicating with the service layer using open interfaces.

Directly related to ETSI TS 102 690 is the ETSI TR 102 898 report [51] regulating the main issues of smart metering architecture, dedicated to automotive applications, in particular in the following aspects: fleet management (which includes tracking stolen vehicles) and communication between vehicles and a fixed infrastructure, the so-called V2I (*Vehicle-to-Infrastructure*). The application scenarios presented there, although adapted to the specificity of the vehicle traffic, are part of the overall structure of M2M systems proposed in [49] (Fig. 2.1).

- TR 102691 [52] is a compilation of six *use cases*, defined as typical for *smart metering* with specification of their elements, including typical beneficiaries, data flow in various modes of operation (normal, emergency, diagnostic, etc.) and failure management: F1, metrology register remote reading and providing results to authorized organizations; F2, two-way communication between a measurement system and authorized organizations; F3, a meter with an advanced billing functionality; F4, a gauge with power on/off functionality; F5, a meter capable of communicating with specific devices in the building (and possibly directly controlling them); and F6, a meter being an interface between the metering system and the portal/gateway allowing to display reports.

2.2 The IEEE P2413 Standard

The task of defining the Internet of Things and its processes, in the most general sense, is dealt with by a working group called 'IoT Architecture' established under the IEEE [54], whose work is still underway, supported by its members including Cisco Systems, Hitachi Ltd., Honeywell International, Huawei, Intel, Kaspersky Lab, Mitsubishi Electric, NIST, Qualcomm Inc., Rockwell Automation, Siemens AG, SigFox, Toshiba Corporation and ZTE. Participation of such a strong representation of shareholders of the ICT market in this initiative is a proof of their broad understanding of the need to regulate its various aspects as early as on the threshold of rapid growth of the IoT devices (i.e. 'things') participating in the network traffic.

The motivation for that undertaking was an observation that a large part of standardization efforts in the area of new telecommunication technologies seems to be vertically structured, which usually leads to the emergence of many isolated technological 'islands' characterized by large (and often unnecessary) redundancy. This verticality consists in developing a solution for a specific need using a 'top-down' method, i.e. from the definition of the physical layer up to the user model, application and business models. Redundancy, on the other hand, is a natural consequence of such an approach, and preventing it consists in avoiding unnecessary multiplication of solutions, achieved by means of using shared standards at different structural levels of systems. The purpose of the IEEE P2413 standard is, therefore, to promote inter-domain interaction. A 'domain' points at a specific application area, i.e. reading from media gauges (electricity, gas, water, etc.) and transporting people and goods, production, trade, building automation, health care, etc. This interaction is believed to enable cooperation between various systems and retain their functional compatibility which, in turn, will facilitate and accelerate the implementation of the IoT if fragmentation of this market into numerous individual solutions is prevented. The intention behind creating this standard is to define a framework architecture of the Internet of Things, including the description of various IoT domains and their abstractions, and to identify common features of these domains. This architecture is to become a reference model defining relationships between different domains (e.g. transport, health services, etc.) and their shared architectural elements. It will also facilitate drawing up certain rules to ensure security and privacy of data sent within the IoT systems. Moreover, such a reference architecture includes the definition of basic architectural blocks of multilayer ICT techniques, so that they can be integrated in the form of multi-tier IoT systems. For the purpose of creating framework architecture, the P2413 working group uses the ISO/IEC/IEEE 42010 [55] standard as a foundation, which major task was – in the face of complexity and abundance of various ICT network organization systems, organization of trade chains, management industrial processes, fleets, employees, etc., with the simultaneous need to communicate between them and keeping other complex relations between them healthy – to develop a general concept of system architecture, including methods of its description and key features related to its behaviour, composition and evolution. These descriptions are necessary for users (organizations, companies, institutions, etc.) so that they can develop, use and manage modern systems capable of communicating smoothly and operating in an integrated, consistent environment. As Fig. 2.2. shows, the ISO/IEC/IEEE 42010 standard is an attempt to reconcile various interests represented by various real domains (related to specific applications) using an abstract IoT domain which is an entity resembling a class or structure, whose components are:

- User/stakeholder – represented by a group of interested recipients.
- Concerns – goals intended to be achieved by the user/shareholder.
- Architectural view – containing main components of a given system, e.g. its hardware and software part of an ICT company.

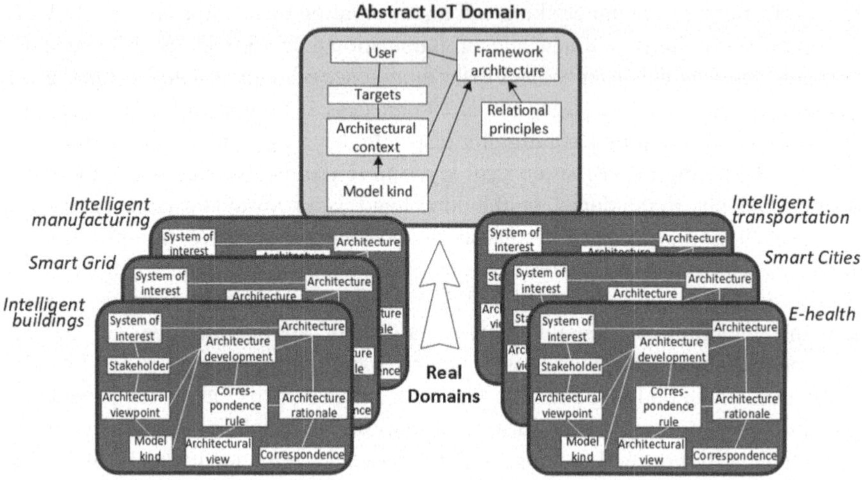

Fig. 2.2 The IEEE P2413 standard architecture framework [54] based on the structure proposed in [55]

- Model kind – its representation, e.g. in the form of flow diagrams, class diagram, Petri network, state machines, etc.
- Correspondence rules – conditions that must exist between subcomponents of a given class in the context of a real domain (e.g. a condition stating that a subcomponent of an architecture context component, called the *operating system*, must be installed on each *computer* subcomponent included in the *hardware* component).
- Architecture framework – a structure describing relationships between system components (defined by the architectural view) and relational principles binding within it, depending on the type of user and the user's goals.

2.3 Standardization According to ETSI: LTN

Another initiative in the IoT standardization was undertaken by ETSI, focusing its activities on a lower level of abstraction than IEEE, with a clear emphasis on specifics, primarily on the practical side of real telecommunication networks dedicated to traffic typical of the Internet of Things. Several assumptions constituting a starting point and, at the same time, main paradigms of the Low-Throughput Networks (LTN) have become a conceptual basis of this initiative [61]:

- Recognizing the *primacy of real radio coverage* over other quality parameters of transmission, such as bandwidth, delay or error rate. This assumption made it possible to classify such systems for a specific group of applications, mainly

related to telemetry, featured by a relatively rare (regular or sporadic) transmission of small volume packets.
- *High-energy link budget* for the needs of increasing the range, compared to the conventional radio communication systems used so far in telemetric applications. As a comparison, this budget, expressed as the difference between EIRP and sensitivity P_{min} or the maximum coupling loss in Eq. (2.1), equals 144 dB for GPRS, 164 dB for NB-IoT or slightly less so in case of other LTN group systems, such as LoRa, SigFox, Weightless-P or Ingenu [62].
- System optimization to support *numerous small volume and low-bandwidth transmissions*. This paradigm additionally favours the concept of simplifying construction of devices, as well as communication protocols and multiaccess mechanisms (which in LTN are contention-free), and extending the device maintenance-free lifetime when battery-operated.
- *Energy saving* which is a sine qua non of long battery life, achievable due to the use of effective sleep mechanisms between transmissions, DC cycle limitations (usually below 1% for terminals or below 10% for base stations), according to Table 2.11, or by limiting EIRP. The two last techniques (i.e. putting to sleep and limiting EIRP) are fostered by adoption of unlicensed bands dedicated to IoT systems, EU433 and EU863–870 (or US 902–928 MHz in the USA), where limits on both EIRP and DC are imposed, as will be discussed in Sect. 2.4. As a reference, normative power consumption expectations for devices representing different families of telecommunication solutions can be given, as presented in Table 2.1 (for comparison with Table 8.8 presenting NB-IoT devices battery lifetime, as the main representative of CIoT family systems), estimated for a traffic model assuming the transmission of 12 100-byte messages per day.
- *System low servicing and maintenance costs.*

Prior to discussing the outcomes of ETSI standardization efforts on networks dedicated to the Internet of Things, it is worth mentioning the M(eter)-Bus EN 13757 (M-Bus) standard, developed for the purposes of remote unified data transmission from meters to recorders (EN13757-1) [56], defining in EN13757- [57] the physical and link layer while specifying the application layer in EN 13757-3 [58].

Table 2.1 Estimated energy consumption of LTN devices compared to other systems

Parameter	Internet of Things systems		Conventional GPRS/3G cellular systems	Typical multi-hop networks, compliant with IEEE 802.15.4 protocol
	LTN	Wireless M-Bus		
Modem power consumption (averaged)	24 μWh only UL[a] 120 μWh UL/DL[a]	75 μWh only UL 80 μWh UL/DL	4000 μWh	150–700 μWh
3 V battery life, assuming 2.4 Ah	20 years	20 years	2 months	2–10 years

[a]UL/DL indicate directions of transmission: 'UL' for uplink, 'DL' for downlink

The last of the substandards, EN 13757-4 [59] (also called Wireless M-Bus), describes the methods of wireless data transmission. Wireless M-Bus can still be included in the WSN networks family, as it is used for communication over relatively short distances, i.e. from dozens (indoors) up to hundreds of meters in open spaces.

In 2014, ETSI standardization groups published three specifications of IoT networks dedicated to low-bandwidth communication [60]. The requirements and recommendations contained there [61–78] are a breakthrough in the M2M connection market, assuming the following: reduction of costs of servicing a given facility (meter, counter, sensor, etc.) down to several Euro annually, few mWatts of transmission power and a modem cost of less than 1 Euro. These assumptions, crucial for the success of the IoT standardization and implementation, were also revolutionary in the context of laying grounds for a wide range of new and innovative applications. Low-Throughput Networks (LTN), described in these specifications, define two-way wireless long-distance networks with a number of features distinguishing them from traditional networks. These networks can transmit data over long distances (up to 40 km over open spaces) or communicate with devices installed in harsh propagation environments, such as wells, basements, manholes or other locations from which radio transmission is significantly impeded due to high suppression of the signal propagation. Moreover, they are supposed to operate at a minimal power consumption, ensuring long-term operation on standard battery supply. In addition, advanced signal processing should provide significant immunity to interference.

Summarizing, the abovementioned features make LTN particularly well-suited for handling M2M connections when data volumes are small, bandwidth requirements low and the delay tolerance high (Fig. 2.3), as is typically expected from

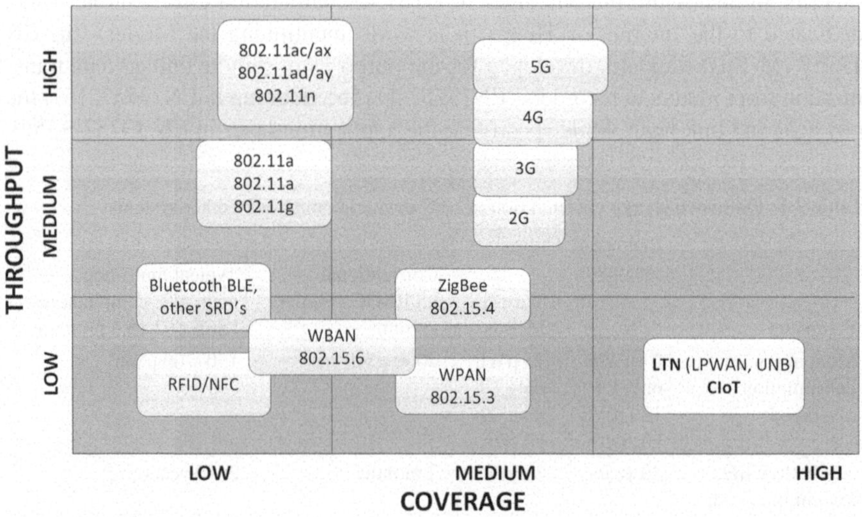

Fig. 2.3 Position of Internet of Things systems (LTN) on the bandwidth/coverage map

telemetric applications (i.e. those for transmission of measurement data), such as smart metering of water, gas or electricity consumption or for the Smart city's needs, including environmental measurements, such as pollution monitoring or urban lighting.

The main challenge for LTN is to provide communication for a myriad of objects participating in the M2M movement and thus constituting the IoT space. The vast majority (especially in mass IoT) requires small capacities, but their operation must be economical both from the standpoint of cost and energy; it must also be scalable. There are some similarities between LTN and traditional wireless data networks (e.g. GSM/GPRS), as concerns, e.g. the channel width – in the range of tens/hundreds of kHz, bit rate – kilobits/sec or adaptive changes in transmitted power. The differences will, however, appear in relation to the requirements for the time of end devices operation without battery replacement and coverage and, thus, the number of base stations necessary to cover the area of a country or the number of supported client devices, which points at the next fundamental feature of these networks, namely, scalability. Low power, low bandwidth, long battery life and simple yet reliable communication mechanisms (e.g. multiple access techniques, acknowledgement methods or ways to provide high processing gain) are the leading topics in the first of the mentioned ETSI specifications – GS LTN 001 (Sect. 2.3.1) – containing a list of various application scenarios. The other two specifications, i.e. GS LTN 002 and GS LTN 003, present, respectively, the functional architecture as well as protocols and interfaces in LTN.

2.3.1 Application Scenarios of the LTN (ETSI GS LTN 001)

LTN have been defined in [61] as telecommunication systems for long-range wireless networks allowing data transmission in a single radio hop over distances up to 40 km in open space or shorter distances, but reaching terminals located in high-pathloss environments (e.g. basements, underground garages, wells, channels, etc.). Moreover, the use of simple modulations (and, hence, low bit rates), effective signal processing methods and redundant coding makes them remarkably immune to noise and interference. These features also make LTN systems particularly well suited for carrying low-throughput and high delay-tolerant M2M traffic. In turn, low duty cycles contribute to low-power consumption, often allowing several short messages per day, or even less (e.g. once a day, per week, etc.), depending on the specificity of the monitored phenomenon. A practical and detailed summary of implementation details specific to various LTN applications can also be found in the ETSI TR 103435 specification [77] that differentiates between a few types of applications:

- Smart cities (Table 2.5)
- Automotive industry (Table 2.6)
- Smart grid, subdivided into options with high (Table 2.7) and low (Table 2.8) measurement expectations

- Intelligent media measurements (also in terms of variants: high, as in Table 2.9, and low, as in Table 2.10)
- Agriculture (plant cultivation, animal husbandry)

The duplex mode, used later in the chapter, referred to as '1.5-way' is intended to express dominant UL transmissions, with DL transmissions taking place only sporadically. The non-integer number '1,5' is used to symbolically emphasize this asymmetry of telecommunication traffic, typical in telemetry. The flagship example of a system with built-in restrictions, both for size of maximum transmission frame (up to 26 bytes) and their allowed number transmitted per day (up to 140), is SigFox, a system described in detail in Sect. 5.3.

Figure 2.4 illustrates the most general reference architecture for the LTN. The nodes directly adjacent to the place of information generation regarding the measured phenomenon (temperature, electricity, amount of gas consumed, etc.) are termed *LTN end points*, or LEP, connected via the radio access segment to one or more *LTN Access Points*, or LAP. Depending on the specific system, the following names can also be used synonymously for a LEP (depending on an IoT system): a User Equipment (UE), a Mobile Station (MS), an End Device (ED) or simply a 'terminal'. The naming used to express a LAP, in turn, includes the following: a base station (BS) or, in cellular networks, a Node B (NB).

The LTN is managed by the back-end system, mediating the service delivery to the customer service platform (CSP). Although the structure itself is not very much different from a typical radio access system, it should be noted that the LTN system network ranges from open user terminal interfaces on the one side to access points,

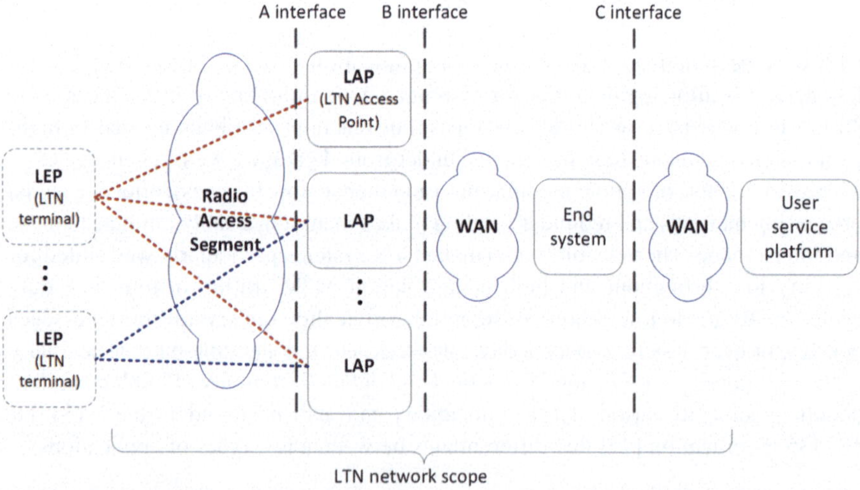

Fig. 2.4 A general LTN organization diagram [61]

Table 2.2 Comparison of LTN transmission frame sizes with other standards

System family	LTN	GSM/ SMS	GPRS	Wireless M-Bus	IEEE 802.15.4 (ZigBee)	PMR
Typical frame size (data field)	12 B[a] (total frame size: 26 B)	140 B	IP packet up to 65 kB	30 –130 B	104 B (total frame size: 127 B)	100 B

[a]'B' denotes a byte

administration systems, long-distance networks and service providers on the other side. From a technical point of view, the key factors distinguishing LTN are as follows:

– Low-radiated power of LTN terminals, forced by the energy-saving requirement, economic construction of the transmitter (including integrated power amplifier with a transmission module, as described in Chap. 8) and legal regulations concerning the frequency bands (as described in Sect. 2.4)
– Very small sizes of transmission frames and low bandwidths
– High link budget, exceeding the budgets of typical data transmission systems, including cellular ones by approx. 20–30 dB
– Low operating costs
– Low maintenance costs

In order to meet these goals, IoT systems should be equipped with specific protocols, transmission frame structures and sizes, acknowledgement methods and power consumption. Although particular solutions may differ, on the background of other known systems, ETSI has stipulated some universal transmission parameters for LTN systems. Typical transmission frame data field sizes, shown in Table 2.2, point at yet another feature of these systems compared to other systems, i.e. a relatively large protocol overhead of up to several dozen percent, measured by the ratio of the number of octets added to the field of user data related to the total frame length. In traditional systems (e.g. those using the TCP/IP), this overhead is minimized by transmitting the longest possible frames providing an acceptable error rate in given propagation conditions in the transmission channel, expressed by the signal-to-noise-and-interference ratio (SNIR). Meanwhile, LTN transmissions of long data packets are naturally ruled out because of the very nature of the telemetric information that by and large involves short messages (in the order of several dozen/several hundred kilobytes) informing about, for example, the status of meters or meteorological situation. The actual load for the radio access segment in these networks originates not so much from the size of individual messages as from their aggregated number arriving from all LEP devices serviced by a single LAP. This number, as already mentioned, may reach hundreds or even thousands of information sources due coverages of a few kilometres in urban conditions and dozens of kilometres in open areas.

Table 2.3 presents other characteristic parameters, allowing to compare LTN systems to other systems. Owing to their reduced bandwidths, LTN prove to be superior with respect to other systems used traditionally in telemetry, which puts them in

Table 2.3 Comparison of LTN main transmission parameters with other standards

System family	LTN	GSM/SMS	GPRS	Wireless M-Bus	IEEE 802.15.4 (ZigBee)
Typical applications	Alarms, telemetry, control, sensor readings, tracking	Alarms, health checks	Telemetry, measurements, control, tracking	Intelligent measurements	Home and process automation, measurements
Distinctive feature	Low-power sensors, interregional compatibility (at the same band), portability, mobility	Capability of communicating directly between devices	Mobility, roaming	Stationary networks	Limited power, dedicated to measurement data transmission
Typical application bytes/day	<200 B/day (typical)	Several SMS/day (140 B)	<10–15 kB/day	<200 B/day	Typically about 150 kB/day
Max. kb/day	<50 kB/day (max.)			<50 kB/day	Max. <1 GB/day
Transmission speed typically per 1 message	<10–100 b/s (typical)	Not applicable	35 kb/s	2.4–4.8 kb/s	250 kb/s @2.4 GHz
Maximum possible speed	<50 kb/s (maximum)			19.2 kb/s	20 kb/s @ 868 MHz
Maximum delays	From milliseconds to seconds	Seconds	640 ms (ping)	From ms to seconds	250 ms @2.4 GHz 786 ms @ 868 MHz (min. 15 ms)
Typical range	Up to 10–12 km (cities)	3 km (cities)	3 km (cities)	Up to 1 km (cities)	<30 meters (inside a building)
	Up to 40–60 km (open spaces)	30 km (open spaces)	30 km (open spaces)	Up to 7 km (open spaces)	
Frequency ranges	<1 GHz ISM: 433–868–915 MHz	900, 1800 MHz	900, 1800 MHz	169 MHz ISM	ISM: 868 MHz, 915 MHz, 2,4GHz

Link budget	Up to 170 dB (@ 100 bps, EIRP = 500 mW)	144 dB (EIRP = 2 W)	144 dB (EIRP = 2 W)	Up to 150 (@2.4 kb/s, 500 mW)	115 dB
Average power consumption per modem	24 μWh (unidirectional)	4000 μWh (general for 2G/3G phones)		75 μWh (unidirectional)	150–700 μWh
	120 μWh (bidirectional)			80 μWh (bidirectional)	
Battery life (maximum theoretical)	10 years	2 months		10 years	2–5 years

Table 2.4 Reference values of transmission parameters related to handling various telemetry tasks

		UL (uplink)			DL (downlink)		
	Number of LEP [million]	Regularity	Frame size [bytes]	Daily load [kbyte]	Regularity	Frame size [bytes]	Daily load [kbyte]
Water meters	30	4/h–1/day	200	7324	1/week	50	262
Gas meters	16	4/h–1/day	100	351,652	1/week	50	262
Electricity meters	57	6/h–1/day	Several hundred		1/week	Several hundred	
Waste management	25	1/h	50	117	No data	No data	No data
Air cleanliness	0.2	1/h	1000	3515	2/day	1000	293
Noise	0.2						
Public lighting	1	1/day	20,000	3906	2/day	1000	390
Parking lot management	4	1/h	100	187,500	1/day	100	7812
City bikes	0.2	4/h	50	2344	1/h	50	586

a leading position in terms of energy efficiency and achieved coverage. As will be demonstrated in Sect. 8.1, the maximum MCL for which cellular IoT (CIoT) systems are designed is 164 dB, which is 20 dB more than in case of GPRS, currently still used as a substitute system to the modern dedicated IoT solutions.

$$\mathrm{MCL}[\mathrm{dB}] = \mathrm{EIRP}[\mathrm{dBW}] - P_{\min}[\mathrm{dBW}] \tag{2.1}$$

In order to better illustrate the amount of data generated by various consumer media covered by the regular reading plan, see Table 2.4 based on [48], a reference quantitative distribution of sensors, meters and various types of meters was presented for a population of 100 million inhabitants. The list also contains typical transmission parameters, important in the LTN planning stage, such as the reporting regularity or transmission frames sizes. These values are only indicative, because as described [48], consumer media (water, gas, electricity) measurement is subject to a wide variety of factors, depending on the region, legal regulations and internal policy of the supplier. It is worth noting that in smart metering applications, an underlying assumption is that the transmission parameters will not change with time; however, the number of sensors is expected to gradually increase. From this perspective, the trend in IoT is different from that observed in wideband data transmission, where exponential growth of transmitted data is caused by both parallel growth in the number of users (information sources) and the growing volume of the generated data.

Table 2.5 LTN applications for the needs of smart cities

Application	Frequency	Data size	Duplex mode
Street parking	1/min to 1/h, depending on traffic Required max. UL delay of ~30 s	Single bytes	1- or 2-directional
Street lighting	UL: 1/day (event log + meter reading); DL: only commands (necessary only in case of faults)	100–200 bytes	2-directional
City bikes rental	1/day to 20/day	1–15 bytes	1-, 1.5- or 2-directional
Trash collection	1/day to 5/day	1–15 bytes	1-, 1.5- or 2-directional
Watering/irrigation	1/day to 5/day	1–15 bytes	1-, 1.5- or 2-directional
Sewage management	1/day to 5/day	1–15 bytes	1-, 1.5- or 2-directional
Flood control	1/day to 5/day	1–15 bytes	1-, 1.5- or 2-directional
Weather monitoring	1/day to 20/day	1–15 bytes	1-, 1.5- or 2-directional
Infrastructure safety (e.g. bridges) monitoring	1/day to 5/day	10–100 bytes	1-, 1.5- or 2-directional
Tracking temporarily released prisoners	1/day to 20/day	1–15 bytes	1-, 1.5- or 2-directional

A practical and detailed summary of implementation details specific to various LTN applications can also be found in the ETSI TR 103435 [77] specification, further subdivided into application:

- Smart cities (Table 2.5)
- Automotive industry (Table 2.6)
- Smart Grid divided into variants: with high (Table 2.7) and low (Table 2.8) measurement requirements
- Smart media measurements, also broken into variants: high (Table 2.9) and low (Table 2.10)
- Agriculture: plant cultivation, animal husbandry

It should be clarified here that the duplex designation of '1.5-directional' is intended to express a distinct dominance of transmission in the UL direction, with DL transmission occurring only sporadically. The non-integer '1,5' is, therefore, used to symbolically mark out this asymmetry of telecommunications traffic, typical of telemetry.

Table 2.6 Automotive applications of the LTN

Application	Frequency	Data size	Duplex mode
Geolocation support	Occasionally	250–1 kbyte	1.5- or 2-directional
Shock detection	Occasionally	10–100 bytes	1-, 1.5- or 2-directional
Driver profiling	1/day to 4/day	250–1 kbyte	1-, 1.5- or 2-directional
Exception reporting	1/week	100–200 bytes	1- or 2-directional
Remote diagnostics/fault reporting	1/week to 1/day	100–750 bytes	1- or 2-directional
Tolls	2/day to 4/h	100–500 bytes	2-directional
Reporting fuel efficiency	1/day	100–500 bytes	1- or 2-directional
Car sharing management	1/h	100–500 bytes	2-directional
Monitoring stolen/ stationery vehicles	1/min to 1/h, depending on traffic. Max UL delay: ~30 s	10–20 bytes	1- or 2-directional

Table 2.7 Smart grid LTN applications in the high-expectation metering variant

Application	Frequency	Data size	Duplex mode
Fault indicator	1/h	3 bytes (typically)	1-directional
Low voltage sensors	4/h	3 bytes (typically)	2-directional
Transformer status monitoring	1/h	10 bytes (typically), 20 bytes (max.)	1-directional
Alarm detection	Once every 2 years	3 bytes (typically)	1.5-directional
Consumption sampling (registration)	1/min to 1/h	No data available	No data available (local archiving)
Readings transmission	6/h to 1/day, periodically	Several hundred bytes + header (optional)	1- or 1.5-directional
DL commands	1/month up to 1/year	5–1 kbyte	1- or 1.5-directional
Software update	1/year to one time throughout the life of the device	Several kbytes	1.5- or 2-directional
Alarm transmission	Occasionally	6–200 bytes	1.5- or 2-directional
Communications testing	1/h to 1/day	Min. 1 byte	1.5- or 2-directional
Real-time clock update	1/day to 1/week	Up to 10 bytes	1- or 1.5-directional
Encryption key update	1/day to 1/year		2-directional
Transmission to a home display[a]	1/h to 1/day	From 10 to 200 bytes	1- or 2-directional
Maintenance	1/day to 1/year	From 10 to a dozen or so kbytes	2-directional
Prepay service management	1/day to 1/month	Several hundred bytes	2-directional

[a]Transmission technique other than LTN: wired is possible (PLC, Ethernet, etc.) or wireless (ZigBee, Bluetooth, etc.)

Table 2.8 Smart grid LTN applications in the low-expectation metering variant

Application	Frequency	Data size	Duplex mode
Consumption sampling = (registration)	1/min to 1/h	No data available	No data available (local archiving)
Readings transmission	6/h to 1/day	Multiple bytes + header (optional)	1- or 1.5-directional
DL commands	1/month up to 1/year	5–1 kbyte	1- or 1.5-directional
Software update	Optional	No data available	No data available
Alarm transmission	Occasionally	6–200 bytes	1.5- or 2-directional
Communications testing	1/h to 1/day	Min. 1 byte	1.5- or 2-directional
Real-time clock update	1/day to 1/week	Up to 10 bytes	1- or 1.5-directional
Encryption key update	Optional	No data available	No data available
Transmission to a home display[a]	1/h to 1/day	From 10 to 200 bytes	1- or 1.5-directional
Maintenance	1/day to 1/year	From 10 bytes to a dozen or so kbytes	1- or 1.5-directional (local)

[a]Transmission technique other than LTN: wired is possible (PLC, Ethernet, etc.) or wireless (ZigBee, Bluetooth, etc.)

Table 2.9 Applications of the LTN for the needs of intelligent media measurements (water, gas) in the high-requirements variant

Application	Frequency	Data size	Duplex mode
Consumption sampling = (registration)	4/h to 1/h	No data available	No data available (local archiving)
Readings transmission	4/day to 1/day, periodically	10 bytes (1 register) up to 200 bytes + header (optional)	1- or 1.5-directional
DL commands	1/month commands up to 1/year	5–50 bytes	1.5- or 2-directional
Software update	1/year to 1 time throughout the device lifetime	Several kbytes	1.5- or 2-directional
Alarm transmission	Occasionally	6–25 bytes	1- or 1.5-directional
Communications testing	1/h to 1/day	Min. 1 byte	1.5- or 2-directional
Real-time clock update	1/day to 1/week	Up to 10 bytes	1.5- or 2-directional
Battery status update	1/day to 1/month	Up to 10 bytes	1- or 1.5-directional
Encryption key update	1/day to 1/year		2-directional
Transmission to a home display[a]	1/h to 1/day	From 10 to 200 bytes	1- or 1.5-directional
Maintenance	1/day to 1/year	From 10 to 1 kbyte	1.5- or 2-directional

[a]Transmission technique other than LTN: wired is possible (PLC, Ethernet, etc.) or wireless (ZigBee, Bluetooth, etc.)

Table 2.10 Applications of the LTN for the needs of intelligent media measurement (water, gas) in the low-requirements variant

Application	Frequency	Data size	Duplex mode
Consumption sampling = (registration)	1/h to 1/day	No data available	No data available (local storage)
Readings transmission	1 / day to 1 / month, periodically	10 bytes (1 register) to 200 bytes + waking preamble (optional)	1- or 1.5-directional
DL commands	1/month commands up to 1/year	5–50 bytes	1.5- or 2-directional
Software update	Optional		1.5- or 2-directional
Alarm transmission	Occasionally	6 to 25 bytes	1- or 1.5-directional
Communications testing	1/day to 1/month	Min. 1 byte	1.5- or 2-directional
Real-time clock update	1/day to 1/month	Up to 10 bytes	1.5- or 2-directional
Battery status update	1/day to 1/month	Up to 10 bytes	1- or 1.5-directional
Encryption key update	Optional		2-directional
Transmission to a home display[a]	1/day (optional)	From 10 to 200 bytes	1- or 1.5-directional
Maintenance	1/month commands up to 1/year	From 10 to 1 kbyte	1.5- or 2-directional

[a]Transmission technique other than LTN: wired is possible (PLC, Ethernet, etc.) or wireless (ZigBee, Bluetooth, etc.)

2.3.2 A Functional Architecture of the LTN (ETSI GS LTN 002)

Functional architecture of the LTN, described in this chapter and presented in Fig. 2.5, bases on the generic network structure shown in Fig. 2.4 [63]. It emphasizes implementation details, in particular the division of competence/tasks in the entire IoT system among its various components, and the flow of information between them, in particular:

- A LEP (LTN terminal) consists of two functional parts: a measurement data acquisition module and an LTN module. Although the interface between them is not standardized, it is recommended that it should be compatible with the AT commands standard or other serial standards, such as SPI, RS232, I2C or equivalent. The LTN module is responsible for receiving data from the acquisition module and transferring it to the radio interface (marked 'A' in Fig. 2.5), its main features being:

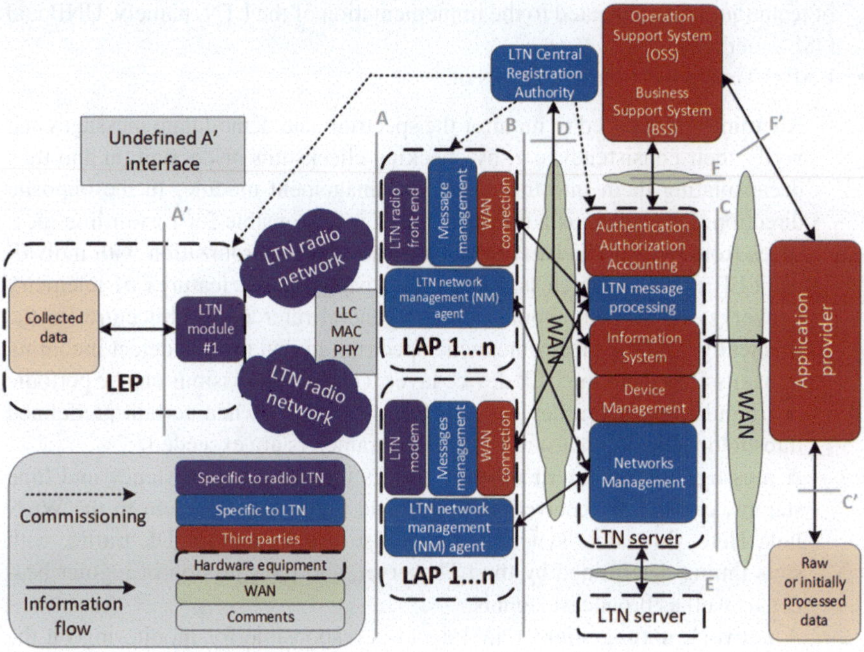

Fig. 2.5 A functional architecture of LTN systems [63]

- It can work without pre-synchronization, which favours low-energy consumption.
- It can function as a transmitter and a receiver.
- In certain applications, it is possible to dynamically change encryption keys and radio parameters.
- Transmission using one of the two techniques (or both at the same time):
  - Spectrum spreading using Orthogonal Sequence Spread Spectrum (OSSS) sequences
  - Ultra-narrowband (UNB) transmission
- Any radio modem, once manufactured, is fully functional and capable of working without the necessity of retrofitting, e.g. with a SIM card.
- Any radio modem has a unique identifier.

- Every message sent from the LEP can be received by several LAPs in the cooperative reception mode. This feature may significantly contribute to improving reception quality, even when there is no synchronization between LAPs.

- **LTN radio network**, feasible in one of the two main transmission techniques, i.e. OSSS or UNB, providing devices on both sides of the link a sensitivity P_{min} usually in the order of −130 dBm, which feature distinguishes them from other systems (as shown in Fig. 2.3), ensuring large coverage. In principle, two groups of techniques are dedicated to the implementation of the LTN, namely, UNB and OSSS methods.

- LAP (LTN Access Point) consists of:
  - **A radio modem** used to monitor the spectrum and demodulate messages and verify their consistency, e.g. by checking checksums or decryption and then decapsulating them into the message management module. In the opposite direction, DL, it acts as a transmitter and is responsible for forwarding messages from the LTN to LEP. Due to the lack of synchronization with individual LEP devices, which is one of the distinguishing features of telemetry networks, the modem should pe most general reference architecture for the rmanently scan the electromagnetic spectrum in real time to detect incoming transmissions from the LEP device layer. These transmissions can be periodic or irregular, e.g. from devices which transmit only when certain predefined threshold values of measured physical parameters are exceeded.
  - **A message management module** checks the format, consistency and time stamps, caches the received messages and then forwards them to the WAN link. Here, time slots synchronization also takes place for DL traffic, with slots timing determined by the LTN server, and transmission of regular beacon as well as broadcast signals.
  - **A network management (NM)** agent is responsible for monitoring on the one hand and configuration functions on the other, such as spectrum observation, radio interface settings, hardware and software status, remote software upgrades and other services operating in LAP.
- **LTN server**, which consists of the following blocks:
  - **Message processing**, responsible for combining messages received in cooperative mode; authentication, authorization and accounting (AAA); transfer of authenticated messages to the information system; management and storage of the LEP traffic database; and determining the LEP location. Optionally, this server may also be responsible for managing the configuration of terminal devices (e.g. transmission speed, transmit power or channel allocation); generation of acknowledgement messages for the terminals; and route updating for downlink traffic and scheduling DL multicast transmissions.
  - **Network management**, endowed with the following functions: LAP authentication; management and monitoring of LAP connections with WAN networks; radio spectrum monitoring; monitoring of LAP components; and software downloads management. All data is then transferred to the Operations Support Systems (OSS).
- **Interfaces:**
  - **Interface A** constituting a radio point of contact of the LTN, using two basic techniques mentioned before, namely, OSSS and UNB, differing significantly in the width of the transmission channel: relatively wide (with BW several

hundred kHz) in the former and very narrow (with BW < kHz) in the latter case [78].

- **Interface B** – a point of contact between LAPs and LTN servers, using a WAN link with IP, such as ADSL, fibre optics, radio links or satellites.
- **Interface C**, between the LTN server and the content provider, based on the IP, analogous to the dIa interface in 'smart metering', as in Sect. 2.1.
- **Interface D**, between the LTN server and the central registration management module, based on the IP.
- **Interface E**, between LTN servers, based on the IP for exchanging information between servers for roaming purposes.
- **Interface F** is a point of contact between LTN servers and OSS-/BSS-based systems (servers), basing on the IP for the purpose of transferring traffic related to the registration process and network status.

Interfaces outside the scope of the ETSI specification:

- **Interface A′**, a point of contact between the measurement system and the LTN/LPWAN radio module
- **Interface C′**, which is the end user system interface from the application provider
- **Interface F′**, based on IP, located between the application provider and the OSS/BSS server, designed to manage the LEP registration and/or network status

Although initial assumptions regarding LTN systems implemented the option of only unidirectional (i.e. just uplink) transmission, the general model assumes two-directional communication. This approach seems to be confirmed in the facts, taking the Weightless-N system as an example of a unidirectional solution (see also Sect. 5.2), which although witnessed several practical implementations (including in London and Copenhagen), but due to the lack of further interest, its development was halted by a group of Weightless SIG companies. The main complaint from potential end users was the lack of the second transmission direction (i.e. downlink), despite satisfactory coverage achieved by the systems (on the order of a few km in urban conditions).

The DL traffic transmission is initiated in the application server which sends a request to the LTN server to forward the message to the most suitable LAP. The LAP, in turn, sends it via the LTN to the recipient (a LEP), as in Fig. 2.6.

In the uplink traffic, in turn, LEP sends messages with or without priority to one or more LAPs via radio packets. Each of the LAPs receiving them verifies their consistency and then transmits them using a secured IP link to the LPWAN server. The server then combines duplicate messages sent from multiple LAPs (base stations); performs authentication, authorization and accounting (AAA); and prepares them for transfer to the application server, which they reach via standard API.

It should be added that, despite functional separation of both main LTN implementation techniques (i.e. UNB and OSSS) at the radio interface level, their interoperability is promoted due to dual modems on the LEP and LAP sides [63]. This creates a certain redundancy which obviously increases the entire system reliability,

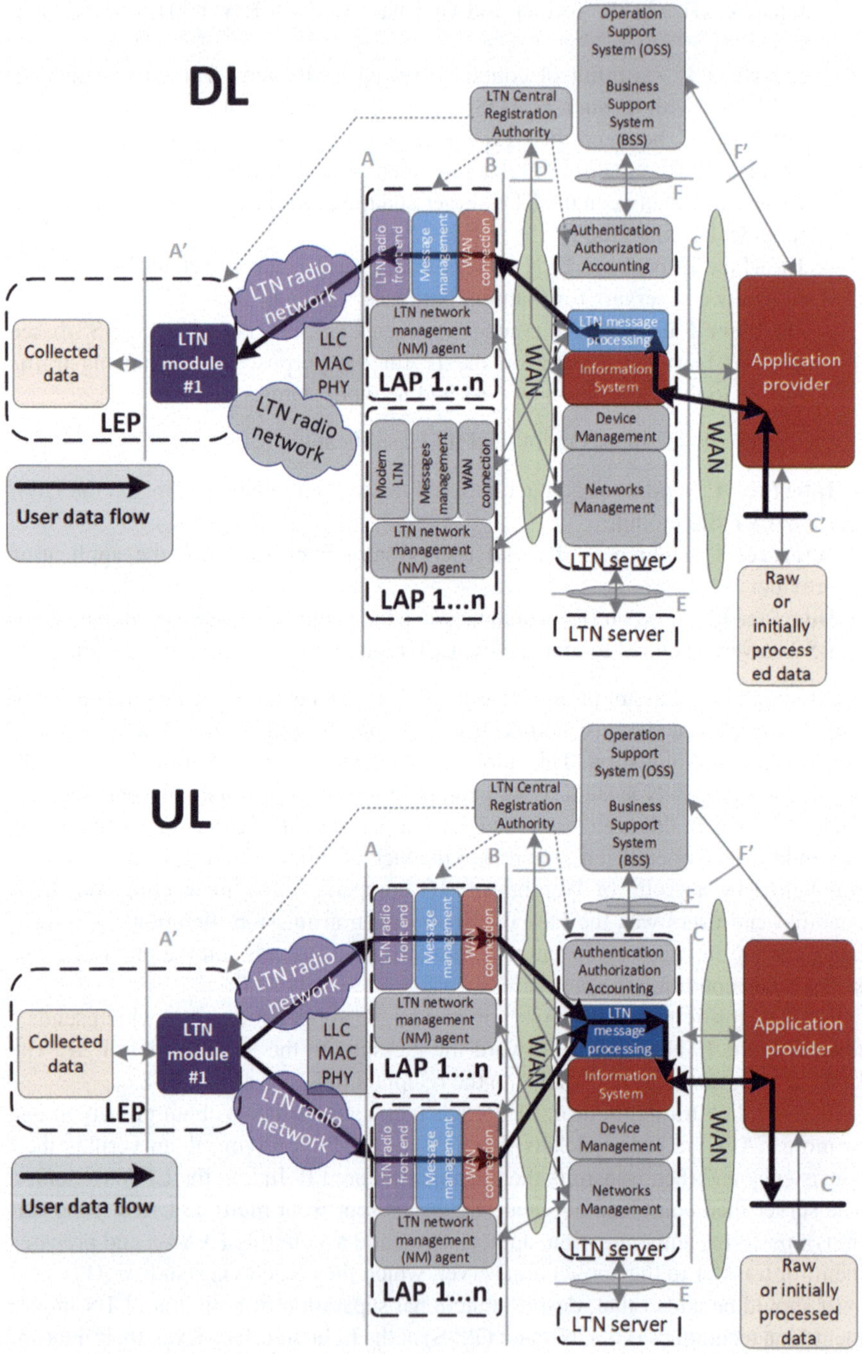

Fig. 2.6 Data flow routes in LTN systems in the following directions: DL and UL [61]

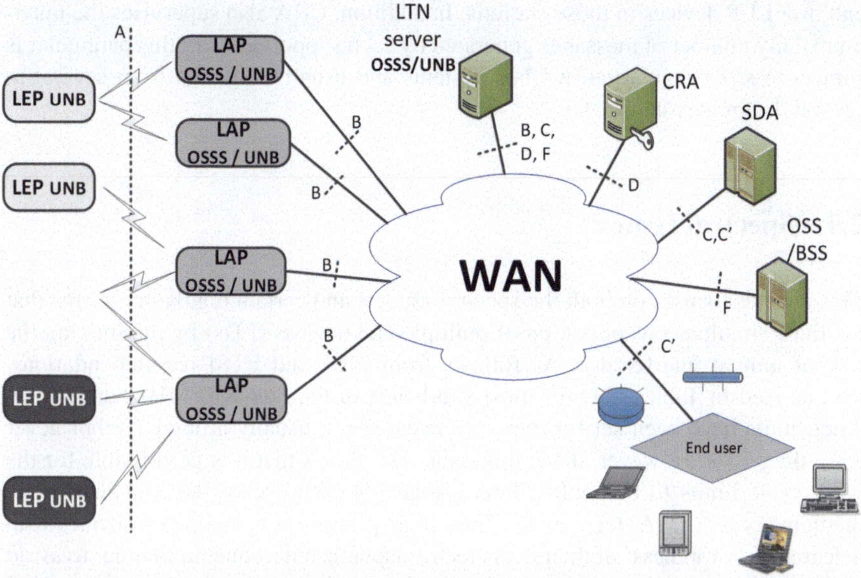

Fig. 2.7 A2 type interoperability between OSSS and UNB LTN [63]

not necessarily entailing an increase in energy expenditure, since individual modems need not to work in parallel. They can share work time by alternately switching on/off radio modules according to a predefined schedule.

Network configurations, taking into account the presence of OSSS- and UNB-type traffic, assume three types of interoperability, with the most advanced type being type A2, allowing for the involvement of the smallest amount of equipment on the LTN:

- Interoperability A1 – a terminal (LEP) contains two modules (UNB, OSSS) which allow communication with LAP supporting any type of LTN. The scenario assumes separate servers for connections from UNB and OSSS networks.
- A2 interoperability – the base station (LAP) contains radio modules representing both systems (UNB, OSSS), so it is able to communicate with the terminals (LEP) using any of these techniques. The scenario assumes one shared OSSS/UNB server, as illustrated in Fig. 2.7.
- Interoperability B1 – similar to the A1 scenario, it assumes that individual LAPs will be dedicated to support only one of the technologies (OSSS or UNB), with the traffic being directed to the shared server.

It is worth noting here that in order for the UNB network to operate correctly, an additional architectural block is needed, namely, a Central Registration Authority (CRA) being a server managing identification codes of devices and base stations, also responsible for generating encryption keys. In practice, the implementation of CRA is the responsibility of the institution or company authorized by the manufacturer of UNB devices, such as SigFox, to install UNB LAP base stations and register end user LEP devices in those stations. In addition, CRA also supervises the maximum daily number of messages generated by each supported LEP; this parameter is subject to strict regulation in UNB systems and usually amounts from several to several dozen messages a day.

2.4 Spectral Issues

This chapter focuses on both the spectral aspects and certain regulatory norms that facilitate simultaneous operation of multiple end devices (EDs) by minimizing the risk of mutual interference. As follows from ERC and ETSI recommendations, summarized in Table 2.11, for most sub-bands in the 868–870 MHz band, fairly strict limits have been set for the work cycle (DC), usually around 1% but never exceeding 10%. However, the comparison also shows that it is permissible for the work cycle limits to be applied interchangeably with one of the so-called polite medium access (PMA) techniques. Their main purpose is to equip end devices with a degree of 'awareness' of their own electromagnetic environment, in order to avoid collisions with transmissions from other end devices. The mechanism used to achieve this goal is the clear channel assessment (CCA) technique (see Sect. 2.5.1) defining two methods of polite access to the medium, AFA (see Sect. 2.5.2) and LBT (see: Sect. 2.5.3), both developed with the intention of a large number of users coexisting in shared frequency bands.

Due to favourable propagation conditions resulting relatively low signal attenuation, IoT systems assume working in bands below 1 GHz, while the requirement of general availability and cost-effective use has limited these bands to the group of unlicensed ISM bands presented in Table 2.11. Basic documents defining these bands are the following: ETSI EN 300220–2 specification [81], ERC recommendation 70–03 [97] and the European Table of Frequency Allocation [82], originally dedicated to, among others for short-range devices (SRD), and adapted also for long-range needs of narrowband IoT systems. These bands by definition allow for narrowband transmission, with significant limitations imposed on the maximum effective radiated power (ERP) and duty cycle (DC). These restrictions are set out in Article 8 (2) of Directive 2014/53 [83], defining the specification of radio interfaces and the allocation of radio equipment classes to them.

Among the bands harmonized in the European Union listed in Table 2.11, particularly noteworthy is band EU873–870 MHz including 'K' – 'R' sub-bands, particularly popular among IoT systems hardware manufacturers. As illustrated in Fig. 2.8, a total of 6.65 MHz is available for use, usually not exceeding 1% of the duty cycle. The band also includes ranges for other systems such as hearing aids and wireless voice systems.

Table 2.11 Main ISM frequency bands (according to [81, 82, 97]) designated for narrowband IoT systems

Band symbol	Frequency range	Max. ERP	Limited connection access	Max. BW	Potential use
D	169.400–169.475 MHz	500 mW	DC ≤ 10%	50 kHz	Assistive listening devices (ALD) short-range devices
E	169.400–169.4875 MHz	10 mW	DC ≤ 0.1%	Entire band	
F	169.4875–169.5875 MHz	10 mW	DC ≤ 0.001% DC ≤ 0.1% between 00:00–06.00 local time	Entire band	
G	169.5875–169.8125	10 mW	DC ≤ 0.1%	Entire band	
H	433.05–434.79 MHz	10 mW	DC = 10%	Entire band	Nonspecific SRD
I	433.05–434.79 MHz	1 mW[a]	–	Entire band	
J.	434.04–434.79 MHz	10 mW	–	25 kHz	
K	863–865 MHz	25 mW	DC ≤ 0.1 or polite access to the medium[b]	Entire band[c]	Microphones, wireless audio systems, e.g. for monitoring children, headphones
L	865–868 MHz	25 mW[d]	DC ≤ 1% or polite access to the medium[b]	Entire band[c]	RFID readers and tags
M	**868–868.60 MHz**	**25 mW**	**DC ≤ 1% or polite access to the medium[b]**	**Entire band[c]**	IEEE 802.15.4 (ZigBee), LoRa, SigFox, Wireless-M, Z-Wave
N	868.70–869.200 MHz	25 mW	DC ≤ 0.1% or polite access to the medium[b]	Entire band[c]	Industrial, Wireless-M
O	869.40–869.65 MHz	25 mW	DC ≤ 0.1% or polite access to the medium[b]	Entire band	Industrial and communications (Wireless-M, LoRa, SigFox)
P	**869.40–869.65 MHz**	**500 mW**	**DC ≤ 10% or polite access to the medium[b]**	**Entire band**	
Q	869.70–870.00 MHz	5 mW	DC no limits	Entire band	Fire and burglar alarm systems
R	869.70–870.00 MHz	25 mW	DC ≤ 1% or polite access to the medium[b]	Entire band	

[a]Additionally for BW > 250 kHz, power spectral density (PSD) should be equal −13 dBm/10 kHz
[b]Implemented using one of the following techniques: LBT or AFA (as explicitly stated in the corresponding table in [97])
[c]Excluding SRD short-range applications limited to BW = 300 kHz
[d]PSD = −4.5 dBm/100 kHz, also possible PSD = 6.2 dBm/100 kHz with limited bandwidth only to the range 865–868

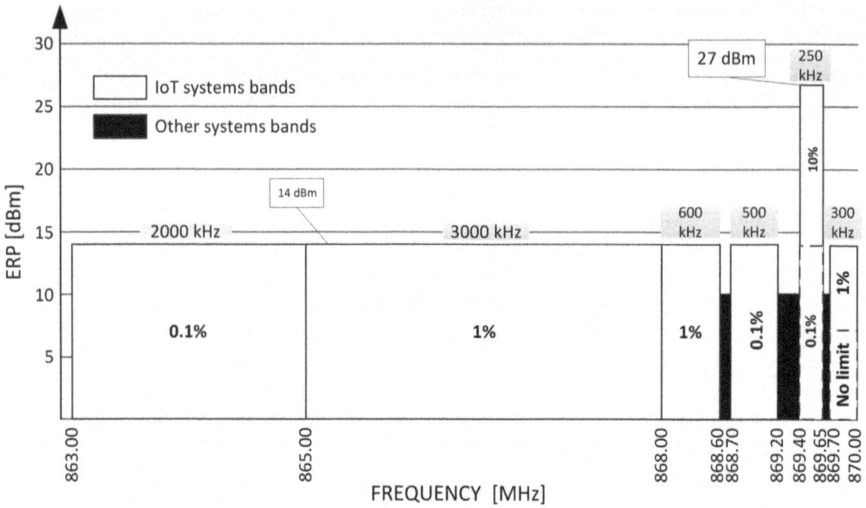

Fig. 2.8 Division of the ISM 863–870 band into sub-bands according to ETSI EN 300 220-2

2.5 Medium Access Techniques in the IoT

2.5.1 The CCA Mechanism

A common element in PMA techniques is the use of the so-called clear channel assessment (CCA), a method also used, for example, in the carrier sense multiple access/collision detection (CSMA/CA) scheme used in WLAN systems [76]. It consists in listening to a specific radio channel to determine if it is unoccupied for a period of time. If the average signal strength level in this period is found to be below a predefined CCA threshold, the device starts transmitting. If, however, the average signal strength level exceeds the threshold value, the device will refrain from transmitting. The next channel listening window, at the same carrier, will not take begin until a certain back-off period has expired. Alternatively, the device can select another operational channel and back off again, and if the channel is found idle, it will commence transmitting. The recommended threshold values for channel occupancy, for the needs of CCA procedures, are presented in Table 2.12.

The sensitivity, P_{min}, on the other hand, is defined as the minimum signal value at the receiver input which, after demodulation, meets one of the following two conditions, understood as general criteria of reception quality assessment. It should not be lower than that resulting from formula (2.2), where BW is the receiver bandwidth, a parameter determined by the manufacturer, usually defined as a 3-dB receiver selectivity. The assessment of reception quality based on these criteria also requires that the measurements be performed with deactivated forward error correction (FEC) and automatic repeat request (ARQ). The aforementioned conditions facilitating the sensitivity P_{min} determination are:

Table 2.12 Thresholds for detecting channel occupancy in CCA procedures according to [80]

Parameter	Value
CCA threshold for devices with ERP <100 mW	15 dB above the level of sensitivity given in the formula (2.2)
CCA threshold for devices with ERP in the range of 100–500 mW	11 dB above the level of sensitivity given in the formula (2.2)

Note: The above limits have been determined for antennas with a gain G = 0 dBd (i.e. +2.15 dBi). For other gain values, the limits should be proportionally rescaled
Example: When using an antenna with gain $G = -5$ dBd, the CCA threshold should be raised by 5 dB

- The bit error rate (BER) of no more than 10^{-3} with no protective coding
- The message success ratio is $(1 - p)^n$, where p is the probability of a single bit error (i.e. 10^{-3}) and n is the number of bits.

$$P_{\min}\left[\text{dBm}\right] = 10\log BW_{\text{kHz}} - 117\text{dB} \tag{2.2}$$

2.5.2 *The AFA Technique*

The Adaptive Frequency Agility (AFA), [81] described in ETSI EN 300 220-1, relies on a device ability to use channel adaptation consisting of regular or event-triggered changes in the following:

- Transmission speed (hence 'adaptation') letting it be adapted to propagation and interference conditions, adjusting the speed so that in certain interference conditions connectivity could be sustained at the cost of bandwidth which by definition is not, as already indicated, a critical parameter in IoT systems. The operation of such an adaptation in conditions of controlled interference and strongly multipath propagation was demonstrated on the example of the LoRa system in Sect. 5.1.
- Working channel (hence 'agility'). Preferable algorithms here ensure an even distribution of data traffic load between available channels while avoiding channels permanently or temporarily occupied by other devices. This, of course, requires that the minimum number of available channels should be two. Thus, the probability of choosing a particular channel from a pool of n available channels is $1/n$.

2.5.3 *The LBT Technique of Polite Medium Access*

The 'Listen Before Talk' (LBT) technique has been described in ETSI TR 102313 [79]. The main LBT principle is simple: a device can transmit at any time when the channel is free. If it is occupied, the device pauses the transmission until the channel has been released or until the frequency has changed to another, unoccupied channel. Such a mechanism, being in fact a kind of a state machine, requires establishing transition time rules (including sleeping, waking, etc.) between states and remaining in certain states, e.g. sleep, transmission, listening, etc. The LBT protocol is therefore characterized by the following parameters recommended for an LTN:

- A minimal transmitter inactivity time (TX_{off}), allowing other users who do not support LBT to gain access to the channel, with the recommended duration $TX_{off} > 100$ ms.
- A minimum receiver listening time t_L, i.e. the time interval in which the device, after having buffered data for transmission, listens to the channel to determine if it is idle. This time must include a listening time constant component t_F and a pseudorandom component t_{PS}, so that the total listening time is given by the Eq. (2.3):

$$t_L = t_F + t_{PS} \tag{2.3}$$

where

- The constant component $t_F \geq 5$ ms.
- The pseudorandom component should be within 0–5 ms in accordance with the assumptions:

  (a) If, at the beginning of the listening period, the channel is free, t_{PS} should be set to zero.
  (b) If, at the beginning of the listening period, the channel is busy, t_{PS} should be hardware-selected from the interval 0–5 ms with 0.5 ms resolution.

- A maximum transmitter activity time during a single TX_{on} transmission. This restriction is intended to prevent a channel from being predominated by transmission from a single device, which makes the required TX_{on} to be as short as possible. On the other hand, in order to successfully communicate with the receiver which is regularly put to sleep for the time RX_{off} for saving its energy, it is necessary to meet the relationship $TX_{on} > RX_{off}$. The maximum transmitter activity time is separately defined for two cases:
  - A single transmission, then $TX_{on} < 1$ s
  - Multiple transmissions, i.e. transmission of either acknowledgements or other devices' polling sequence, then $TX_{on} < 4$ s

- A maximum time of awaiting an acknowledgement (ACK). Despite the lack of formal recommendations regarding this parameter, it should be noted that if the beginning of the confirmation has not been received before the end of the constant component t_F (=5 ms), the channel may be occupied by another transmitter.

Moreover, the receiver should be fairly immune to 'blocking' by the presence of strong adjacent channels.

Chapter 3
Propagation Models for LTN Systems

According to the frequency bands set out in Sect. 2.4, propagation loss estimation models should cover the 169 MHz to 1 GHz range, which reduces the available pool of known empirical models, depending on (f_c) the channel centre frequency, to the following:

- For the range 150 MHz $\le f_c \le$ 450 MHz: Hata model [64, 65]
- For the range 450 MHz $< f_c \le$ 1000 MHz: WINNER+ model [66] in the extended version with the 450–2000 MHz range, based on the original WINNER II model [67] covering the 2–6 GHz frequency bands
- For the range $f_c \le$ 1000 MHz: a simple 3GPP model based on the 3GPP TR 45.820 ([45] Annex D, Table D.1)
- For the range 0.8 GHz $< f_c \le$ 100 GHz (or up to 30 GHz for the RMa variant): an extended 3GPP model defined in the 3GPP TR 38.900 [98] specification

3.1 The Hata Model

This model is a mathematical representation of the original attenuation curves obtained by Okumura for Tokyo in the frequency range of 100–3000 MHz [72]. Algorithmic approximation of these curves in the Hata model, on the other hand, covers the range of 150–1500 MHz; the other parameters of the model were also limited, including the orographic profile of the area narrowed to smooth and almost smooth areas. The model which calculates the attenuation median (L_{50}) is defined in the formula (3.1), where:

- f_c – centre frequency of the channel [MHz]
- h_t – height of the base station antenna (BS) [m] above the ground
- h_r – height of the user terminal antenna (UE) [m] above the ground
- d – distance [km] between the transmitter and the receiver
- the designation ‘log(x)’ means ‘$\log_{10}(x)$’

K. Staniec, *Radio Interfaces in the Internet of Things Systems*,
https://doi.org/10.1007/978-3-030-44846-2_3

The model can be used in the following parametric range:

- $1 \leq h_r \leq 10$ m
- $30 \leq h_t \leq 200$ m
- $1 \leq d \leq 20$ km
- 150 MHz $\leq f_c \leq$ 1500 MHz (here: narrowed down to the 150–450 MHz band)

3.1.1 Urban Areas

$$L_{50}[\text{dB}](\text{urban}) = 69.55 + 26.16\log(f_c) - 13.82\log(h_t) + (44.9 - 6.55\log(h_t)\cdot\log(d)) - a(h_r) \tag{3.1}$$

$a(h_{\text{mob}})$ denotes the correction coefficient of the terminal antenna height calculated according to dependence (3.2) and (3.3):

- For small and medium towns (and at $1 \leq h_r \leq 10$ m):

$$a(h_r)[\text{dB}] = (1.1\cdot\log(f_c) - 0.7)\cdot h_r - (1.56\cdot\log(f_c) - 0.8) \tag{3.2}$$

- For large towns:

$$a(h_r)[\text{dB}] = \begin{cases} 8.29(\log(1.54h_r))^2 - 1.1 & f_c \leq 200\,\text{MHz} \\ 3.2(\log(11.75h_r))^2 - 4.97 & f_c \geq 400\,\text{MHz} \end{cases} \tag{3.3}$$

3.1.2 Suburban Areas

$$L_{50}[\text{dB}] = L_{50}(\text{urban}) - 2\left[\log\left(\frac{f_c}{28}\right)\right]^2 - 5.4 \tag{3.4}$$

3.1.3 Open Areas

$$L_{50}[\text{dB}] = L_{50}(\text{urban}) - 4.78(\log f_c)^2 + 18.33\log f_c - 40.94 \tag{3.5}$$

Comment: in case of semiopen areas, 5 dB should be added to the L_{50} values obtained for open areas.

3.2 WINNER+ Model

This model, specified in [66] and extending the applicability of the WINNER II model [67] to the 450–2000 MHz frequency range (of which the list presented here refers to the range up to 1000 MHz), covers several application scenarios, of which for the needs of narrowband IoT systems, and has been distinguished, according to Table 3.1:

- UMi – urban microcell
- O2Ia – macrocell outdoor-to-indoor
- SMa – suburban macrocell
- Uma – urban macrocell
- O2Ib – microcell outdoor-to-indoor

3.3 A Simple 3GPP Model

The model was developed in 3GPP TR 45.820 document ([45] Annex D, Table 3.1) for the needs of cellular IoT systems (see also Chap. 8) and expressed in the empirical formula (3.6), for the reference conditions presented in Table 3.2.

$$L_{ext} = L_0 + 37.6 \cdot \log(d) \tag{3.6}$$

3.4 An Extended 3GPP Model

The full 3GPP model defined in the following documents, 3GPP TR 38.900 [98] and ETSI TR 138 900 V14.3.1, has been elaborated mainly for the fifth-generation cellular systems utilizing frequency ranges above 6 GHz. Similar to the WINNER+ model described in Sect. 3.2, it includes a few scenarios typical to cellular systems, as given in Table 3.3. Its applicability in IoT systems, in turn, is mandated by the fact that according to the source documentation, its pathloss formulas are valid for 800 MHz and above, which covers the sub-gigahertz range recommendable for LTN systems.

3.4.1 *Pathloss*

Formulas for the pathloss calculation are presented in Table 3.3, with distinction between LOS and NLOS variants for five basic scenarios, assuming log-normal fading with a standard deviation of σ:

Table 3.1 WINNER+ model environmental variants (application scenarios)

Scenario		Total attenuation [dB]	Shadow fading [dB]	Applicability range (default height of antenna installation)
Umi (B1)	LOS	$L = 22.7 \log(d) + 27.0 + 20.0 \log(f_c)$	$\sigma = 3$	$10\ \text{m} < d < d'_{BP}$[a] $h_{BS} = 10$ m, $h_{MS} = 1.5$ m
		$L = 40.0\log(d) + 7.56 - 17.2\log(h'_{BS}) - 17.3\log(h'_{MS}) + 2.7\log(f_c)$		$d'_{BP} < d < 5$ km
	NLOS	f_c: 450–1500 MHz $L = (44.9 - 6.55\log(h_{BS}))\log(d) + 5.83\log(h_{BS}) + 16.33 + 26.16\log(f_c)$	$\sigma = 4$	$10\ \text{m} < d < 2000$ m $h_{BS} = 10$ m $h_{MS} = 1.5$ m
O2Ia (B4)	NLOS	When propagation between the base station antenna and the external wall (section L_{ext})[b] occurs at LOS: $L = L_b(d_{out} + d_{ind}) + 21.04 + 14(1 - 1.8 \log(f_c)) + 0.5d_{ind}$ where L_b means L for the UMi (B1) LOS scenario	$\sigma = 7$	$3\ \text{m} < d_{out} + d_{ind} < 1000$ m $h_{BS} = 25$ m $h_{MS} = 3(n_{Fl}-1) + 1.5$[b] 10 m < d < 2000 m
		When propagation between the base station antenna and the external wall (section L_{ext}) occurs at NLOS: $L = L_b(d_{out} + d_{ind}) + 21.04 + 14(1 - 1.8 \log(f_c)) + 0.5d_{ind} - 0.8h_{MS}$ where L_b means L for UMi (B1) NLOS scenario		
SMa (C1)	LOS	$L = 23.8 \log(d) + 27.2 + 20.0 \log(f_c)$	$\sigma = 4$	$30\ \text{m} < d < d_{BP}$[c] $h_{BS} = 25$ m, $h_{MS} = 1.5$ m
		$L = 23.8 \log(d) + 27.2 + 20.0 \log(f_c)$	$\sigma = 6$	$d'_{BP} < d < 5$ km $h_{BS} = 25$ m, $h_{MS} = 1.5$ m
	NLOS	f_c: 450–1500 MHz $L = (44.9 - 6.55\log(h_{BS}))\log(d) + 5.83\log(h_{BS}) + 13.33 + 26.16\log(f_c)$	$\sigma = 8$	$50\ \text{m} < d < 5$ km $h_{BS} = 25$ m, $h_{MS} = 1.5$ m

UMa (C2)	LOS	$L = 26.0 \log(d) + 25.0 + 20.0 \log(f_c)$	$\sigma = 4$	$10\text{ m} < d < d'_{BP}$[a] $h_{BS} = 25$ m, $h_{MS} = 1.5$ m
		$L = 40.0\log(d) + 9.27 - 14.0\log(h'_{BS}) - 14.0\log(h'_{MS}) + 6.0\log(f_c)$	$\sigma = 6$	$d'_{BP} < d < 5$ km $h_{BS} = 25$ m, $h_{MS} = 1.5$ m
	NLOS	f_c: 450–1500 MHz $L = (44.9 - 6.55\log(h_{BS}))\log(d) + 5.83\log(h_{BS}) + +16.33 + 26.16\log(f_c)$	$\sigma = 8$	$10\text{ m} < d < 5$ km $h_{BS} = 25$ m, $h_{MS} = 1.5$ m
O2Ia (C4)	NLOS	When propagation between the base station antenna and the external wall (section L_{ext}) occurs at LOS: $L = L_b(d_{out} + d_{in}) + 21.04 + 14(1 - 1.8 \log(f_c)) + 0.5d_{in}$ where L_b means L for the UMa (C2) LOS scenario	$\sigma = 10$	$10\text{ m} < d < 5$ km $h_{BS} = 25$ m $h_{MS} = 3(n_{FI}-1) + 15$
		When propagation between the base station antenna and the external wall (section L_{ext}) occurs at NLOS: $L = L_b(d_{out} + d_{in}) + 21.04 + 14(1 - 1.8 \log(f_c)) + 0.5d_{in} - 0.8h_{MS}$ where L_b means L for UMa (C2) NLOS scenario		

[a]Breakpoint distance $d_{BP} = 4h'_{BS} \cdot h'_{MS} \cdot f_c / c$, where f_c means frequency in [Hz], $c = 3.0 \cdot 10^8$ m/s means the speed of light in a vacuum, while h'_{BS} and h'_{MS} are effective heights of antennas in the base station and the terminal equipment, respectively. The values are calculated using the formulas: $h'_{BS} = h_{BS} - 1.0$ m and $h'_{MS} = h_{MS} - 1.0$ m, where h_{BS} and h_{MS} are the actual heights of the antennas; the effective height of urban environments is set at 1.0 m

[b]External distance d_{ext} outside indicates the distance between the device located outside the building and the point on the outer wall, the nearest to the device located inside the building. θ is the angle between the external path propagating the signal and the straight perpendicular to the wall. Parameter n_{FI} defines the number of the floor (assuming for the ground floor the value of parameter $n_{FI} = 1$)

[c]Breakpoint distance $d_{BP} = 4h_{BS} \cdot h_{MS} \cdot f_c / c$

Table 3.2 Assumptions for the 3GPP TR 45.820 propagation model

Parameter	Assumed value
Cellular system	Hexagonal mesh, three sectors per cell
Frequency band	900 MHz
Distance between cells	1732 m
Mobile stations speed	0 km/h
User distribution	Uniform on the cell surface
Base station (BS) EIRP per 200 kHz channel	43 dBm
Reference attenuation	L_0 = 120.9 dB for 900 MHz frequency
Standard deviation σ_{ext}	8 dB
BS antenna gain	18 dBi
MS antenna gain	−4 dBi
Feed line losses L_{fld}	3 dB

- RMa – rural macrocell
- UMa – urban macrocell
- UMi-street canyon – urban microcell
- InH-office – indoor home
- InH – shopping mall

The total pathloss in the composite situation, i.e. one where the signal is partially propagated outdoors and partially indoors, is given by the generic formula (3.7), in which L_{ext} represents the external attenuation defined in Table 3.3 for RMa and UMi scenarios, L_{ew} stands for the external wall attenuation, and L_{ind} is the indoor loss dependent on the penetration depth, whereas σ_P is the indoor loss standard deviation. Respective values of coefficients included in the total pathloss L can be found in Table 3.4:

$$L[\mathrm{dB}] = L_{ext} + L_{ew} + L_{ind} + N\left(0, \sigma_P^2\right) \tag{3.7}$$

Signal losses associated with some typical constructional materials are stated in Table 3.5:

The last variant of mixed propagation between different environments is that of the into-vehicle pathloss expressed by formula (3.8), where L_{ext} stands for the outdoor attenuation as given in Table 3.3 for RMa and UMi scenarios, whereas the normal distribution values representing the signal temporal variability are equal to: $\mu = 9$, $\sigma_P = 5$. Optionally, for vehicles with metalized window panes, $\mu = 20$ can be used. The frequency applicability range of this model has been stated as 0.6–60 GHz.

$$L[\mathrm{dB}] = L_{ext} + N\left(\mu, \sigma_P^2\right) \tag{3.8}$$

Table 3.3 Median pathloss in the extended 3GPP model

Scenario		Median pathloss [dB]	Shadow fading [dB]	Applicability range (default antenna height)
RMa	LOS	$L = \begin{cases} L_1 & 10\,\mathrm{m} \le d_{2D} \le d_{BP} \\ L_2 & d_{BP} \le d_{2D} \le 10\,\mathrm{km} \end{cases}$ a,b $L_1 = 20.0\log(4\pi d_{3D} f_c d) + \min(0.03h^{1.72};10)\log(d_{3D}) -$ ef $\min(0.04h^{1.72};14.77) + 0.002\log(h)d_{3D}$ $L_2 = L_1(d_{BP}) + 20\log(d_{3D}/d_{BP})$	$\sigma = 4$ (for L_1) $\sigma = 4$ (for L_2)	$h_{BS} = 35$ m, $h_{MS} = 1.5$ m $W^c = 20$ m, $h^d = 5$ m 5 m $\le h \le$ 50 m 5 m $\le W \le$ 50 m 10 m $\le h_{BS} \le$ 150 m 1 m $\le h_{MS} \le$ 10 m
	NLOS	$L = \max(L_{RMa\_LOS}; L'_{RMa\_NLOS})$ *dla* $10\ \mathrm{m} \le d_{2D} \le 5\ \mathrm{km}$ $L'_{RMa\_NLOS} = 161.04 - 7.1\log(W) + 7.5\log(h)$ $-(24.37 - 3.7(h/h_{BS})^2)\log(h_{BS})$ $+(43.42 - 3.1\log(h_{BS}))(\log(d_{3D}) - 3)$ $+20\log(f_c) - (3.2(\log(11.75h_{MS}))^2 - 4.97)$	$\sigma = 8$	

(continued)

Table 3.3 (continued)

Scenario		Median pathloss [dB]	Shadow fading [dB]	Applicability range (default antenna height)
UMa	LOS	$L = \begin{cases} L_1 & 10\,\text{m} \le d_{2D} \le d'_{BP} \\ L_2 & d'_{BP} \le d_{2D} \le 5\,\text{km} \end{cases}$ g $L_1 = 32.4 + 20 \log (d_{3D}) + 20 \log (f_c)$ $L_2 = 32.4 + 40\log(d_{3D}) + 20\log(f_c) - 10\log\left((d'_{BP})^2 + (h_{BS} - h_{MS})^2\right)$	$\sigma = 4$	$1.5\ \text{m} \le h_{MS} \le 22.5\ \text{m}$ $h_{BS} = 25\ \text{m}$
	NLOS	$L = \max\left(L_{UMa\_LOS}; L'_{UMa\_NLOS}\right) dla\, 10\ \text{m} \le d_{2D} \le 5\ \text{km}$ $L'_{UMa\_NLOS} = 13.54 + 39.08\log(d_{3D}) + 20\log(f_c) - 0.6(h_{MS} - 1.5)$	$\sigma = 6$	
		Optional: $L = 32.4 + 20\log(d_{3D}) + 30\log(f_c)$	$\sigma = 7.8$	
UMi – street canyon	LOS	$L = \begin{cases} L_1 & 10\,\text{m} \le d_{2D} \le d'_{BP} \\ L_2 & d'_{BP} \le d_{2D} \le 5\,\text{km} \end{cases}$ $L_1 = 32.4 + 20 \log (d_{3D}) + 20 \log (f_c)$ $L_2 = 32.4 + 40\log(d_{3D}) + 20\log(f_c) - 9.5\log\left((d'_{BP})^2 + (h_{BS} - h_{MS})^2\right)$	$\sigma = 4$	$1.5\ \text{m} \le h_{MS} \le 22.5\ \text{m}$ $h_{BS} = 10\ \text{m}$
	NLOS	$L = \max\left(L_{UMi\_LOS}; L'_{UMi\_NLOS}\right) \text{for}\, 10\ \text{m} \le d_{2D} \le 5\ \text{km}$ $L'_{UMi\_NLOS} = 35.3\log(d_{3D}) + 22.4 + 21.3\log(f_c) - 0.3(h_{MS} - 1.5)$	$\sigma = 7.82$	
		Optionally: $L = 32.4 + 31.9 \log (d_{3D}) + 20 \log (f_c)$	$\sigma = 8.2$	

InH – office	LOS	$L = 32.4 + 17.3 \log(d_{3D}) + 20 \log(f_c)$	$\sigma = 3$	$1.0\ \text{m} \le d_{3D} \le 100\ \text{m}$
	NLOS	$L = \max\left(L_{InH\_LOS}; L'_{InH\_NLOS}\right)$ $L'_{InH\_NLOS} = 38.3\log(d_{3D}) + 17.3 + 24.9\log(f_c)$	$\sigma = 8.03$	$1.0\ \text{m} \le d_{3D} \le 86\ \text{m}$
		Optionally: $L = 32.4 + 31.9 \log(d_{3D}) + 20 \log(f_c)$	$\sigma = 8.29$	
InH – shopping mall	LOS	$L = 32.4 + 17.3 \log(d_{3D}) + 20 \log(f_c)$	$\sigma = 2$	$1.0\ \text{m} \le d_{3D} \le 150\ \text{m}$

[a]Breakpoint distance d_{BP} identical as in Table 3.1, normalized to 1 m

[b]d_{2D} – total distance between BS and MS, expressed in horizontal projection, normalized to 1 m

[c]W – average buildings width

[d]h – average street height

[e]d_{3D} – real 3D distance between BS and MS, normalized to 1 m

[f]f_c – center frequency normalized to 1 GHz

[g]d'_{BP} – modified breakpoint distance. Modified antenna heights h'_{BS} and h'_{MS} are calculated with the following formulas: $h'_{BS} = h_{BS} - h_E$ and $h'_{MS} = h_{MS} - h_E$, where h_{BS} and h_{MS} correspond to real antenna heights, whereas h_E is an effective urban environment height. For UMi $h_E = 1$ m. For UMa $h_E = 1$ m with the probability $1/(1 + C(d_{2D};h_{MS}))$, in all other cases, it is a uniformly distributed value $(12,15,\ldots,(h_{MS}-1.5))$

$C(d_{2D};h_{MS})$ is given by the formula:

$$C(d_{2D};h_{MS}) = \begin{cases} 0, & h_{MS} < 13\,\text{m} \\ \left(\dfrac{h_{MS}-13}{10}\right)^{1.5} g(d_{2D}), & 13\,\text{m} \le h_{MS} \le 23\,\text{m} \end{cases}$$

where:

$$g(d_{2D}) = \begin{cases} 0, & d_{2D} \le 18\,\text{m} \\ \dfrac{5}{4}\left(\dfrac{d_{2D}}{100}\right)^{3} \exp\left(\dfrac{-d_{2D}}{150}\right), & 18\,\text{m} < d_{2D}. \end{cases}$$

Table 3.4 Warianty środowiskowe (scenariusze aplikacyjne) w rozszerzonym modelu 3GPP

Model type	External wall attenuation L_{ew} [dB]	Indoor loss L_{ind} [dB]	Standard deviation σ_P
High-loss model[a]	$5-10\log\left(0.3\cdot 10^{\frac{-L_{glass}}{10}}+0.7\cdot 10^{\frac{-L_{concrete}}{10}}\right)$	$0.5 \cdot d_{2D\_in}$	4.4
Low-loss model[b]	$5-10\log\left(0.7\cdot 10^{\frac{-L_{glass}}{10}}+0.3\cdot 10^{\frac{-L_{concrete}}{10}}\right)$		6.5

[a]The model applicability range scenarios: UMa, UMi – street canyon, RMa. Most adequate for traditional civil buildings

[b]The model applicability range scenarios: UMa, UMi – street canyon. Most adequate for commercial (e.g. shopping malls) and institutional (e.g. modern office centres) buildings

Table 3.5 Constructional materials penetration losses in the extended 3GPP model

Material	Penetration losses [dB]
Standard multi-pane glass	$L_{glass} = 2 + 0.2f_c$
Infrared reflecting glass (IRR)	$L_{glass\_IRR} = 23 + 0.3f_c$
Concrete	$L_{concrete} = 5 + 4f_c$
Wood	$L_{wood} = 4.85 + 0.12f_c$

Notice: The centre frequency f_c expressed in [GHz]

3.4.2 LOS Conditions Probability

The extended 3GPP model enables to calculate the pathloss under mixed LOS/NLOS conditions, making a distinction between these two cases in a probabilistic manner. As is presented in Table 3.6, the model defines for each of scenarios the probability of line-of-sight conditions P_{LOS}, complementary to which is the probability of non-LOS conditions, hence $P_{NLOS} = 1\text{-}P_{LOS}$. Thus, the resultant propagation attenuation is a combination of both L_{LOS} and L_{NLOS}, in which the probabilities P_{LOS} and P_{NLOS} serve as weighting coefficients, as given by Eq. (3.9).

$$L[\mathrm{dB}] = P_{LOS}L_{LOS} + (1-P_{LOS})L_{NLOS} \tag{3.9}$$

3.5 Calculation of Additional Losses

3.5.1 Losses on Scattering Obstacles

The models described in the previous sections serve, according to the nomenclature adopted in Sect. 3.3, for calculating external attenuation (L_{ext}), i.e. propagation losses between the base station antenna and the external building wall. In addition

Table 3.6 Environmental variants (application scenarios) in the extended 3GPP model

Scenario	LOS conditions probability (distance in [m])
RMa	$P_{LOS} = \begin{cases} 1, & d_{2D} \le 10\,\mathrm{m} \\ \exp\left(-\frac{d_{2D}-10}{1000}\right), & 10\,\mathrm{m} < d_{2D} \end{cases}$
UMi – street canyon	For outdoor users: $P_{LOS} = \begin{cases} 1, & d_{2D} \le 18\,\mathrm{m} \\ \frac{18}{d_{2D}} + \exp\left(-\frac{d_{2D}}{36}\right)\left(1-\frac{18}{d_{2D}}\right), & 18\,\mathrm{m} < d_{2D} \end{cases}$ For indoor users: Formula as above with d_{2D} replaced with d_{2D_out} being the horizontally projected distance between BS and the building corner
UMa	For outdoor users: $P_{LOS} = \begin{cases} 1, & d_{2D} \le 18\,\mathrm{m} \\ \left(\frac{18}{d_{2D}} + \exp\left(-\frac{d_{2D}}{63}\right)\left(1-\frac{18}{d_{2D}}\right)\right)\left(1 + C'(h_{MS})\frac{5}{4}\left(\frac{d_{2D}}{100}\right)^3 \exp\left(-\frac{d_{2D}}{150}\right)\right), & 18\,\mathrm{m} < d_{2D} \end{cases}$ Gdzie: $C'(h_{MS}) = \begin{cases} 0, & h_{MS} \le 13\,\mathrm{m} \\ \left(\frac{h_{MS}-13}{10}\right)^{1.5}, & 13\,\mathrm{m} < h_{MS} < 23\,\mathrm{m} \end{cases}$ For indoor users: Formula as above with d_{2D} replaced with d_{2D_out} being the horizontally projected distance between BS and the building corner
Indoors – office type	$P_{LOS} = \begin{cases} 1, & d_{2D} \le 1.2\,\mathrm{m} \\ \exp\left(-\frac{d_{2D}-1.2}{4.7}\right), & 1.2\,\mathrm{m} < d_{2D} < 6.5\,\mathrm{m} \\ \exp\left(-\frac{d_{2D}-6.5}{32.6}\right)\cdot 0.32, & 6.5\,\mathrm{m} \le d_{2D} \end{cases}$
Indoors – open space	$P_{LOS} = \begin{cases} 1, & d_{2D} \le 5\,\mathrm{m} \\ \exp\left(-\frac{d_{2D}-5}{70.8}\right), & 5\,\mathrm{m} < d_{2D} \le 49\,\mathrm{m} \\ \exp\left(-\frac{d_{2D}-49}{211.7}\right)\cdot 0.54, & 49\,\mathrm{m} < d_{2D} \end{cases}$

Notice: The probability of LOS conditions was derived for the following antenna heights: 3 m for indoors, 10 m for UMi and 25 m for Uma

Table 3.7 Default values of the effective H_{clut} ambient scattering height

Scattering environment type	H_{clut} [m]	Calculation method L_{scat}
Water/sea	10	Equation (3.12)
Open/rural area	10	Equation (3.12)
Suburban area	10	Equation (3.11)
Urban/woodland area	15	Equation (3.11)
Highly urbanized area	20	Equation (3.11)

to thus estimated L_{ext}, additional attenuation associated with clutter, i.e. obstacles in the immediate vicinity of the building should be added, L_{scat}. Document ITU-R P.2108-0 [69], taking into account a variety of clutter around the receiver, is devoted to assessment of this type of extra attenuation. Because the resulting suppression is the sum of various propagation phenomena, which the propagating wave experiences in interaction with these objects (e.g. reflection, refraction, diffraction, scattering), taking into account the shares of particular mechanisms was impractical. The method proposed in the quoted ITU recommendation, therefore, assumes a division into two cases of scattering environment: woodland and built-up areas. It also assumes that diffraction is the dominant propagation mechanism. Regarding other environmental categories, it is assumed that reflection and scattering are dominant phenomena.

L_{scat} losses should be added to free space losses L_{bf} (or L_{ext} calculated in accordance with the procedures in Sects. 3.1 and 3.2), in this situation reducing the actual height of the antennas (h_{TR}): transmission and reception by effective H_{clut} height of the scattering environment.

L_{ext} losses, due to the assumption of diffraction as a dominant phenomenon, are based on the formula (3.10) borrowed from the ITU-R P.526-8 standard, valid for the Fresnel coefficient range of $v \geq -0.78$ (for smaller values assuming $J(v) = 0$) and for situations in which either the transmitter or receiver is below H_{clut}. Unless precise data on the effective height of the scattering environment are available (H_{clut}), one should use values from Table 3.7. Proper attenuation of L_{scat} should be, depending on the case indicated in Table 3.7, calculated using the formula (3.11) or (3.12).

$$J(v)[\mathrm{dB}] = 6.9 + 20\log\sqrt{(v-1)^2 + 1} + v - 0.1 \tag{3.10}$$

$$L_{scat}[\mathrm{dB}] = J(v) - 6.03 \tag{3.11}$$

$$L_{scat}[\mathrm{dB}] = -K_{h2}\log(h_{TR} / H_{clut}) \tag{3.12}$$

where

- $v = K_{nu}\sqrt{h_\Delta \cdot \theta_{rozp}}$.
- $h_\Delta = H_{clut} - h_{TR}$.
- $\theta_{rozp} = \tan^{-1}\left(h_\Delta / w_s\right)[^\circ]$.

- $K_{h2} = 21.8 + 6.2 \log (f)$.
- $K_{nu} = 0.342\sqrt{f}$.
- f is a frequency in [GHz] in the range of 0.03–3 GHz.
- w_s is the street width (27 m by default).

3.5.2 Building Penetration Losses: A 3GPP Model

The model was developed in document 3GPP TR 45.820 ([45] Annex D.1). The total attenuation associated with the EM wave penetration through the structure of the building is given by Eq. (3.13).

$$L_{\mathrm{add}}\left[\mathrm{dB}\right] = L_{\mathrm{ew}} + \max\left(W_{\mathrm{i}} \cdot p; \alpha^{\mathrm{d}}\right) - n \cdot G_{n} \tag{3.13}$$

where

- L_{ew} the loss of penetration through the external wall, drawn with the assumption of a uniform distribution from one of the three ranges of values: 4–11, 11–19 and 19–23 dB.
- W_i is attenuation brought by the internal walls of the building, drawn with the assumption of a uniform distribution in the range of 4–10 dB.
- p is the number of penetrated walls, assuming one of the values from the set {0, 1, 2, 3}, where the case p=3 refers to the UE being under the strongest propagation attenuation, e.g. in basements.
- α is a attenuation exponent equal to 0.6 dB/m.
- d is the depth of penetration in [m], randomized with the assumption of a uniform distribution in the range of 1–15 m.
- G_n is the height gain per floor of the building, equal to 1.5 dB/floor.
- n is the number of floors in the building from the set of values {0, 1, 2, 3, 4}, drawn with the assumption of a uniform distribution.

Two scenarios were adopted for the planning reference for the large-scale design: they differ in percentage distribution of the UE within the three L_{ew} attenuation ranges, the number of EUs placed in locations with extremely high propagation attenuation (p=3) and the type of internal walls in terms of attenuation. The parametric definitions of both cases are given in Tables 3.8 and 3.9.

3.5.3 Building Penetration Losses: WINNER+ and ITU-R Models

Apart from defining the L_{ext} calculation method, the WINNER+ model offers, moreover, the possibility to estimate the expected attenuation also for building interiors (L_{ind}), which coincides with a typical application scenario for IoT, in which signal

Table 3.8 Scenario No. 1: definition of propagation losses parameters

Propagation attenuation distribution for external walls L_{ew}			
Attenuation L_{ew}	4–11 dB	11–19 dB	19–23 dB
Percentage of evenly distributed UE in the given range	25%	65%	10%
Assumptions regarding additional losses related to internal walls penetration			
Percentage of UE experiencing extreme suppression (p=3). The remaining UE distributed evenly within p=0,1,2	15%		
An assumption regarding the dependence of penetration losses on interior walls of the building	Independently, i.e. for each inner wall, a different value of W_i		

Table 3.9 Scenario No. 2: definition of propagation losses parameters

Propagation attenuation distribution for external walls L_{ew}			
Attenuation L_{ew}	4–11 dB	11–19 dB	19–23 dB
Percentage of evenly distributed UE in the given range	25%	50%	25%
Assumptions regarding additional losses related to internal walls penetration			
Percentage of UE experiencing extreme suppression (p=3). The remaining UE distributed evenly within p=0,1,2	20%		
An assumption regarding the dependence of penetration losses on interior walls of the building	Dependent, i.e. one W_i value is drawnassumed to be representative for all internal walls		

sources – counters, meters, sensors, etc. – are often placed in environments that are difficult to propagate, including closed rooms or basements. For channels with $f_c \leq 450$ MHz, for practical purposes, a value of 15–25 dB can be assumed, corresponding to the results of measurements carried out in several residential homes in the UK for selected frequencies in the 88–5800 MHz [68] range. These results were obtained for reception terminals located in the rooms farthest away from the signal source outside the building.

Similar values for the VHF range (30–300 MHz) were obtained in [71] (Table 1.3), for the so-called Class B propagation scenarios, defined as the reception in the interiors of buildings, using a portable terminal, located at a height of 1.5 m above the floor level of the ground floor in a room with a window in the outer wall. For this class, due to the large variation in the results of the measured attenuation associated with the penetration of the signal to the building, a further division of this class into two subclasses, B1 and B2, described below, was made. For thus defined scenarios, the following values of attenuation L_{ind} and standard deviation σ_{dod} have been obtained for the VHF band:

- B1 class: the case of the so-called light interior with a terminal located near a window, surrounded by walls characterized by low attenuation (such as plasterboard walls, thin brick walls, etc.): $L_{\text{ind}} = 9$ dB, $\sigma_{\text{ind}} = 4.5$ dB

- B2 class: the case of the terminal placed deep inside the building, away from the windows, surrounded by more strongly attenuation (thicker) walls: $L_{ind} = 15$ dB, $\sigma_{ind} = 5.0$ dB

Compilation of theoretical and measurement knowledge on the subject of building entry loss is document ITU-R P.2346-2 [73] containing mainly the results of extensive measurements of propagation attenuation during EM penetration into buildings, carried out in many places around the world for 80 MHz to 73 GHz. These results also lead to an interesting conclusion: in terms of the impact on the final loss of penetration, buildings should be arranged in two basic groups with a partition line marked by their thermal efficiency:

- Modern buildings, thermally efficient, containing metallized glass and foiled panels – usually showing a significantly higher attenuation for the penetrating wave. Such buildings have been considered to have thermal transmittances 'U' in the range < 0.3 and < 0.9, respectively, for the entire structure and for windows. In order for a building to be classified as thermally efficient, both its structure and windows should have the value of $U < 1$. If this parameter does not meet this ratio for any of these two components, then the whole building should be considered traditional.
- Traditional buildings, deprived of such materials, thus causing less attenuation.

The final conclusions regarding bands of interest to the IoT systems confirm the validity of the previously quoted (after [71]) median values of L_{ind}:

- For the frequency $f_c = 200$ MHz: $L_{ind} = 9$ dB, $\sigma_{ind} = 3$ dB
- For the frequency $f_c = 600$ MHz: $L_{ind} = 11$ dB, $\sigma_{ind} = 6$ dB

ITU-R P.2346-2 also contains a recommendation regarding the method of calculating the total standard deviation from the median L_{ind} value, which is a combination of deviations for the case of mixed propagation (i.e. external and internal): outside σ_{out} while its measuring range has been determined as 5.5 dB for VHF bands and UHF, as well as an in the building σ_{ind} with the value as above. Due to the lack of a statistical correlation between both propagation paths, the total deviation of σ_{ind} is given in the formula (3.14). According to this formula, in the VHF bands, where $\sigma_{ext} = 5.5$ dB a $\sigma_{ind} = 3$ dB, the total deviation $\sigma_{tot} = 6.3$ dB, while in the UHF bands, where $\sigma_{ext} = 5.5$ dB a $\sigma_{ind} = 6$ dB, the total deviation $\sigma_{tot} = 8.1$ dB

$$\sigma_{tot}\left[\mathrm{dB}\right] = \sqrt{\sigma_{ext}^2 + \sigma_{ind}^2} \tag{3.14}$$

The ITU-R P.2346-2 recommendations, on the other hand, constituted the basis for the building entry loss (BEL) composite model of attenuation associated with the penetration of electromagnetic waves in buildings, valid for the frequency of ~80 MHz to 100 GHz, proposed in ITU-R P.2109-0 [75]. In this model, the main division between thermally efficient and traditional buildings is preserved, with total losses expressed by Eq. (3.15) with the coefficients for the formula given in Table 3.10. An example of a median attenuation chart (i.e. for $p = 0.5$) is presented

Table 3.10 BEL model coefficients (ITU-R P.2109)

Regards →	Median L_{dod} (μ_1)			σ_1		μ_2		σ_2	
Type of building ↓	r	s	t	u	v	w	x	y	z
Traditional	12.64	3.72	0.96	9.6	2.0	9.1	−3.0	4.5	−2.0
Thermally efficient	28.19	−3.00	8.48	13.5	3.8	27.8	−2.9	9.4	−2.1

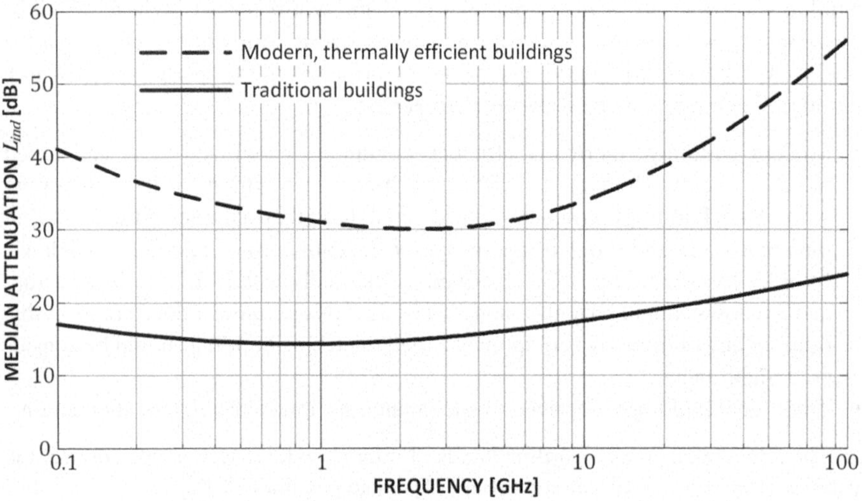

Fig. 3.1 Median building penetration attenuation according to the BEL model (ITU-R P.2109)

in Fig. 3.1. The assumptions and the code of the L_{ind} attenuation program according to the BEL model (ITU-R P.2109) in the MATLAB program are presented in Appendix No. 1 (Chap. 10).

$$L_{ind}\left[\mathrm{dB}\right] = 10\log\left(10^{0.1A(p)} + 10^{0.1B(p)} + 10^{0.1C}\right) \tag{3.15}$$

where

- $A(p) = F^{-1}(p)\sigma_1 + \mu_1$; $B(p) = F^{-1}(p)\sigma_2 + \mu_2$; $C = -3.0$
- $\mu_1 = L_h + L_e$; $\mu_2 = w + x \log(f)$
- $\sigma_1 = u + v \log(f)$; $\sigma_2 = y + z \log(f)$
- $L_h = r + s \log(f) + t(\log(f))^2$ - median losses for perpendicular incidence on the building
- $L_e = 0.212 \bullet |\theta|$ – correction for a different than 0° (perpendicular) angle of incidence with respect to the building
- f – frequency in [GHz]
- θ – the waveform angle with respect to the building (0° corresponds to the perpendicular incidence in L_h) [deg]
- p – probability that the value of losses will be exceeded ($0 \leq p \leq 1$)

- $F^{-1}(p)$ – inverse distribution function of the logarithmic normal distribution as a function of the probability p

Concluding the subject of propagation loss calculations, ITU-R P.2040 should also be mentioned [74], in which the physics of interaction of electromagnetic waves and matter is presented in detail, subject to meticulous analysis of various phenomena in this process, including reflection, scattering, penetration through multilayered materials, etc. while providing tabulated values of electromagnetic parameters of various building materials over wide frequency ranges.

Chapter 4
Radio Interfaces in LTN Networks

As mentioned in Sect. 2.3.2, the LTN group IoT networks can be implemented using one of two basic techniques, namely, OSSS direct spread or UNB (ultra-narrowband) transmission. Regardless of the choice, the most important goal is to obtain a specific G_p processing gain, substantial enough to provide the largest possible tolerance range for the maximum coupling loss (MCL) given by formula (2.1). The MCL, in turn, translates to either of the two effects:

- The system's capability to provide coverage of several or even several dozen kilometres, way above that available with traditional systems (e.g. GPRS).
- Good radio signal penetration to places which are difficult to reach, e.g. through layers of earth and concrete separating the water or gas meter which, in traditional systems, is often a barrier difficult to overcome, as illustrated in Fig. 4.1. Because this type of scenario is typical for IoT applications, transmission techniques used in Internet of Things systems should have adequate mechanisms to compensate for the effect of additional attenuation caused by the mentioned obstacles to propagation, expressed as L_{add} in Fig. 4.1, most often in the range of 20–30 dB.

In the link energy budget of IoT systems, these additional losses are compensated by the processing gain G_p obtained by means of, e.g. the use of spread spectrum, multiple repetitions or narrowing the channel bandwidth (which decreases the noise power contained in them).

Thus, the signal-to-noise ratio, and, in the general case, also signal-to-interference-plus-noise ratio (SNIR), is represented by the dependence (4.1), in which the useful signal strength (S) and the sum of noise (N) and interference (I) are expressed by Eqs. (4.2) and (4.3), respectively, while NF is the noise figure. The 174 dBm/Hz constant in the Eq. (4.3) is the spectral density of noise (N_0) and is the product of the Boltzmann constant ($k = 1.38 \cdot 10^{-23}$ J/K) and room temperature $T = 293$ K. A recommended unit for expressing bandwidth in LTN systems is kHz, due to spectral limits imposed by regulators for bands most commonly used by these systems (i.e. ISM 159 MHz, ISM 315 MHz w USA, ISM 433 MHz, ISM 868/915 MHz). Further in

K. Staniec, *Radio Interfaces in the Internet of Things Systems*,
https://doi.org/10.1007/978-3-030-44846-2_4

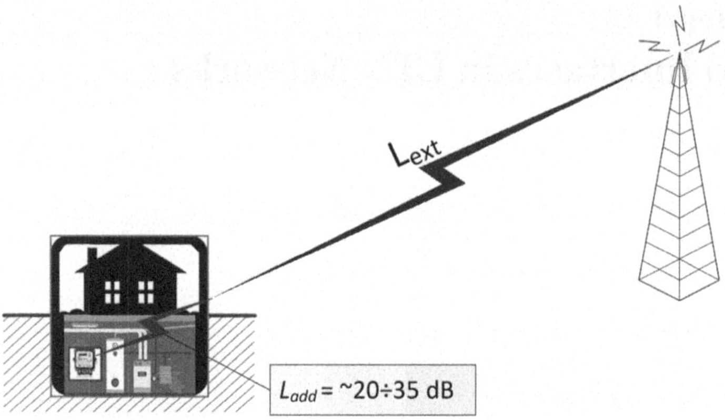

Fig. 4.1 Illustration of additional attenuation introduced by a difficult propagation environment

this chapter, the characteristics of OSSS and UNB systems will be presented, which will help to understand the source of processing gain obtained.

$$\mathrm{SNIR}[\mathrm{dB}] = S[\mathrm{dBm}] - (N + I)[\mathrm{dBm}] \tag{4.1}$$

$$S[\mathrm{dBm}] = \mathrm{EIRP}[\mathrm{dBm}] - L_{\mathrm{ext}}[\mathrm{dB}] - L_{\mathrm{add}}[\mathrm{dB}] - NF[\mathrm{dB}] + G_{\mathrm{Rx}}[\mathrm{dBi}] \tag{4.2}$$

$$N[\mathrm{dBm}] = -174\,\mathrm{dBm/Hz} + 10\log\left(BW[\mathrm{kHz}]\right) + 30\,\mathrm{dB} \tag{4.3}$$

4.1 Spread Spectrum Techniques Used in IoT Systems

4.1.1 Direct Sequence Spread Spectrum (DSSS)

One of the most popular variants of the spread spectrum technique is direct sequence (DS), in which the spreading effect is obtained by performing eXclusive OR (XOR) operation between the information stream (in bits/s) and a fast-changing pseudorandom sequence (in chips/s). In some systems (e.g., UMTS), these sequences are often mutually orthogonal, which enables distinguishing signals from different transmitters. These sequences, along with the idea of spreading in time, are depicted in Fig. 4.2, while its translation into the spectrum domain along with the illustration of the resulting processing gain can be found in Fig. 4.3. Directly proportional relationship between the bit duration – T_b (or chip duration – T_c) and the power spectral density $S(f)$ is evident in those sequences (here on the example of rectangular signals). The observed difference in the power density $S(f = 0)$ between the two versions of the signal, i.e. the despread signal, occupying the BW_{inf} width and the spread

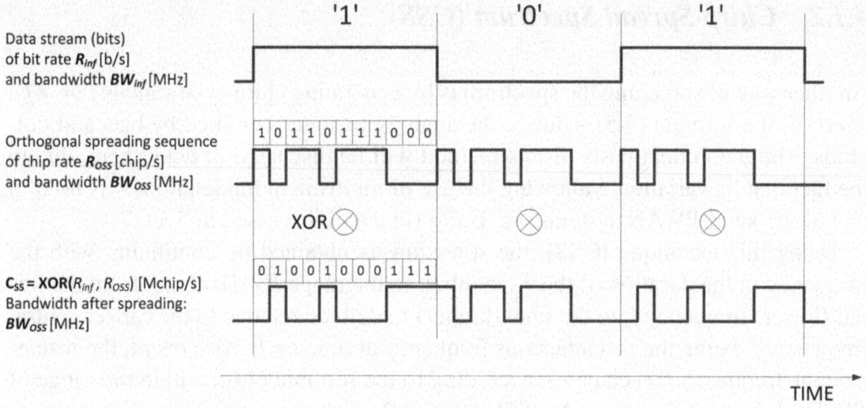

Fig. 4.2 Illustration of an XOR operation performed between a bit stream (R_{inf}) and a pseudorandom spreading sequence (R_{OSS}), in the DSSS method

signal occupying BW_{SS}, helps to formulate the principle defining G_{p}, in accordance with formula (4.4). In order to emphasize the generality of this principle, also appropriate for other SS methods, such as CSS (chirp spread spectrum), time-hopping spread spectrum (THSS) or frequency-hopping spread spectrum (FHSS), 'O' was omitted in the symbol index of the spread channel width ('SS').

$$G_{\text{p}}[\text{dB}] = 10\log\left({BW_{\text{SS}}}\big/{BW_{\text{inf}}} \right) \tag{4.4}$$

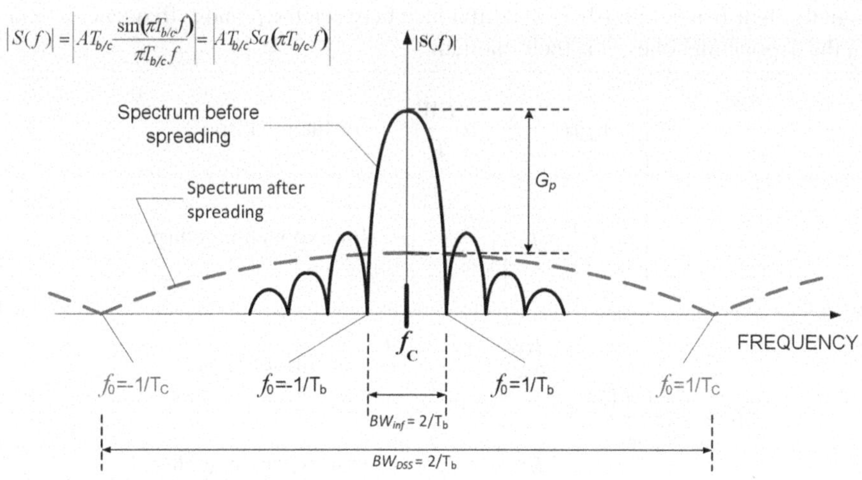

Fig. 4.3 The concept of G_{p} processing gain obtained by the spectral scattering method

4.1.2 Chirp Spread Spectrum (CSS)

Another way of spreading the spectrum is by generating chirp-type signals [84–87], – $c(t)$ in the formula (4.5) – due to the analogy to sounds emitted by bats and dolphins. The theoretical basis of this method will be discussed in detail here, due to the fact that its variation – allowing the use of multivalent modulations – is used in one of the key LPWAN systems, i.e. LoRa (in detail discussed in Sect. 7.1).

Using this technique (CSS), the spreading is obtained by continuing with the frequency in the duration of the T_s, symbol, in the range BW [Hz], i.e. from the initial (lower) frequency f_0 to the final (higher) f_1, defined relative to the carrier central frequency f_c being the instantaneous frequency at time $t = 0$. As a result, the instantaneous frequency $f_i(t)$ changes, according to the formula (4.6), within the range of BW [Hz], in the time interval $-T_s/2 \leq t \leq T_s/2$, with the change of either linear or exponential type.

$$c(t) = \begin{cases} \exp\left(j\varphi(t)\right) & \text{for} -\frac{T_s}{2} \leq t \leq \frac{T_s}{2} \\ 0 & \text{for all other } t \end{cases} \tag{4.5}$$

The parameter μ in this formula takes on a value '+1' (*downchirp*) or '−1' (*upchirp*), which corresponds to a pulse with a rising or falling frequency. In the basic CSS implementation, the value of μ corresponds to the value of a particular bit; usually $\mu = +1$ will correspond to the logical one, and $\mu = -1$ will correspond to the logical zero, although this remains a matter of convention adopted in a given telecommunication system. The parameter α, in turn, determines the rate of frequency changes and is given in the formula (4.7). It shows that in the chirp linear signals, their bandwidth BW is the difference between the f_1 and f_0 frequencies, and in the exponential ones, it is their quotient.

$$f_i(t) = \begin{cases} f_c + \mu\alpha t = f_c + \mu \frac{BW}{T_s} t & \text{for linear chirps} \\ f_c \cdot \alpha^{\mu t} = f_c \cdot \left[\left(\frac{f_1}{f_0}\right)^{\frac{1}{T_s}}\right]^{\mu t} & \text{for exponential chirps} \end{cases} \tag{4.6}$$

$$\alpha = \frac{1}{2\pi}\frac{d^2\varphi(t)}{dt^2} = \begin{cases} \frac{BW}{T_s} = \frac{(f_1 - f_0)}{T_s} & \text{for linear chirps} \\ BW^{\frac{1}{T_s}} = \left(\frac{f_1}{f_0}\right)^{\frac{1}{T_s}} & \text{for exponential chirps} \end{cases} \tag{4.7}$$

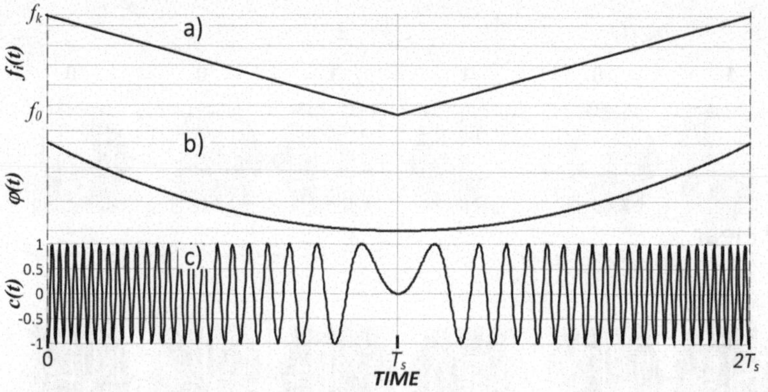

Fig. 4.4 Increasing chirp signal and its processing: (**a**) changes in instantaneous frequency $f_i(t)$; (**b**) phase changes $\varphi(t)$; (**c**) chirp signal in the time domain

Finally, by integrating the instantaneous frequency f_i in formula (4.6), according to the relationship (4.8), one arrives at the dependency expressing the phase of the *chirp* signal, as shown in formula (4.9). It is worth noting that while in $c(t)$ with exponential frequency change the phase also has an exponential character, in signals with linear changes of instantaneous frequency (as in Fig. 4.4a, c), the phase has the character of a quadratic function (Fig. 4.4b).

$$f_{\mathrm{i}}(t) = \frac{1}{2\pi}\frac{d\varphi(t)}{dt} \tag{4.8}$$

Additionally, in order to limit the transmission spectral width, each chirp pulse is smoothed in the time domain by multiplying it by a window function with a raised cosine characteristic [88], given by formula (4.10). The resulting time-domain signal, for example, a binary modulation sequence {1;0;1;1;0;0}, is illustrated in Fig. 4.5.

$$\varphi(t) = \begin{cases} 2\pi\left(f_{\mathrm{c}}t + \mu\frac{\alpha}{2}t^2\right) + \varphi_0 & \text{for linear chirps} \\ 2\pi f_{\mathrm{c}}\left(\frac{\alpha^{\mu t} - 1}{\mu\ln(\alpha)}\right) + \varphi_0 & \text{for exponential chirps} \end{cases} \tag{4.9}$$

$$P_{\mathrm{RC}}(t) = \begin{cases} 1 & \text{for } |t| \le \frac{1-\alpha}{1+\alpha}\frac{T_{\mathrm{s}}}{2} \\ \frac{1}{2}\left[1 + \cos\left(\frac{(1+\alpha)\pi}{\alpha T_{\mathrm{s}}}\left(|t| - \frac{1-\alpha}{1+\alpha}\frac{T_{\mathrm{s}}}{2}\right)\right)\right] & \text{for } \frac{1-\alpha}{1+\alpha}\frac{T_{\mathrm{s}}}{2} \;\#\;\# \le \frac{T_{\mathrm{s}}}{2} \\ 0 & \text{for } |t| > \frac{T_{\mathrm{s}}}{2} \end{cases}$$

(4.10)

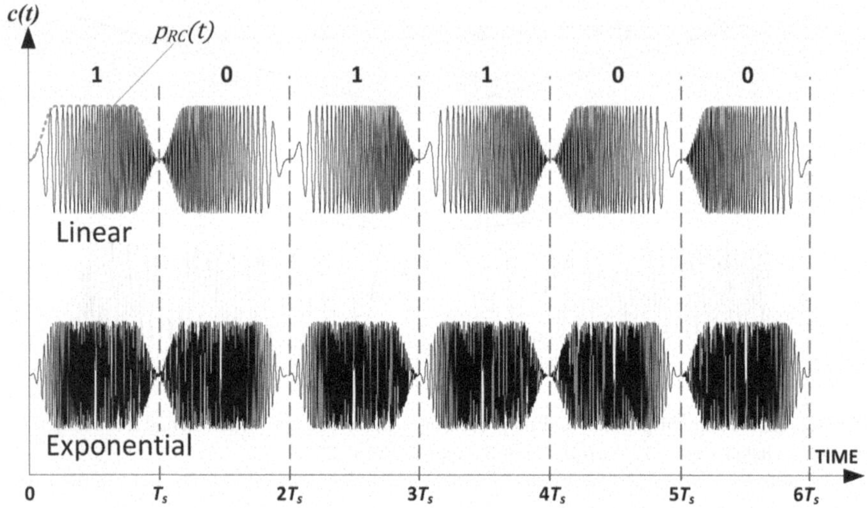

Fig. 4.5 A sequence of pulse chirps in an example of a modulating binary sequence {1;0;1;1;0;0}

On the receiving side, the following properties are used when detecting the signal:

1. The result of multiplying the frequency-increasing chirp signal (*upchirp*) with another signal of increasing frequency is also a chirp signal with the increasing frequency, because the instantaneous frequencies add up (similarly to the mutual multiplication of *downchirp* signals);
2. Multiplication of an increasing chirp signal with a chirp signal with a falling frequency results in a narrowband signal with a fixed frequency, equal to twice the central frequency f_c and unaltered parameters: α and BW. In systems using CSS, signals identical in terms of amplitude, BW and f_c, differing only in the direction of frequency changes (i.e. having the opposite μ) are referred to as coupled signals.

Using these properties, detection of the $c(t)$ signal on the receiving side is possible with the use of a matched filter with an impulse response $s(t)$ being a coupled version of $c(t)$, i.e. a signal with an opposite direction of frequency change. The result of the convolution operation $c(t){*}s(t) = AC(\tau)$ is, in fact, a function of autocorrelation of the $s(t)$ [90] signal, with a high value for the zeroth sample (a beginning of the signal) and negligible for the remaining samples, according to the formula (4.11) for autocorrelation AC. This knowledge will be crucial when discussing the LoRa system, where the sample position τ for which the $AC(\tau)$ function reaches the peak value will also indicate the value of the modulating symbol (see Sect. 5.1.5.1).

$$AC(\tau) = \sqrt{BW \cdot T_s} \frac{\sin\left[\pi BW \cdot t\left(1 - \frac{t}{T_s}\right)\right]}{\pi BW \cdot t} \cos(2\pi f_c t) \quad (4.11)$$

It should be noted that, similarly to all spread spectrum techniques, the CSS method is characterized by low transmission efficiency, expressed in spectral efficiency η defined by the formula (4.12), where R_b is the transmission rate in the physical layer and BW is the bandwidth occupied by the signal $c(t)$, or f_1–f_0, which implies low transmission rates, in favour of high processing gain. This is in line with the previously presented LPWAN system priorities, according to which the range and immunity to interference are more important than R_b bandwidth. The immunity in CSS is caused by the fact that the receiver, in order to detect the value of the transmitted bit, needs only to determine the direction of change (i.e. down/up) of the frequency, after obtaining the time synchronization. What it means is that the very shape of the received signal, even subjected to extensive distortions (e.g. caused by multipath propagation), does not affect the correct character of detection.

$$\eta = {R_b}/{BW} \, [\mathrm{b/s/Hz}] \quad (4.12)$$

The classic CSS technique, discussed in this chapter, maps a single bit to one chirp signal. This information will be helpful in understanding the LoRa system (Sect. 5.1), in which the technique has been modified to allow up to 2048-state modulations obtained by manipulating the trajectory of discrete instantaneous frequency changes.

4.2 IoT Systems with Spread Spectrum (OSSS): General Characteristics

This section is not going to be a presentation of specific LPWAN network solutions, but rather a demonstration of typical radio parameters recommended for such systems [78], adapted to those priority objectives for IoT telecommunications solutions presented in Sect. 2.3. Quoting just the general guidelines is supported by the fact that new LPWAN family systems may be expected in the future in addition to the currently dominant ones (e.g. LoRa, SigFox, Weightless), because in this area of application, the dynamics of progress and often quite nonstandard needs of some users may create stimuli for the development of their own systems, with time adopted and disseminated, and thus recognized as another de facto standards in the IoT market. In addition, it can already be observed that IoT solutions – just as other systems in the past – will not necessarily compete with each other but rather seek to find optimal application niches for themselves. Bluetooth and WLAN, for example, have gone a similar path. Following a short period of competition in the late 1990s

Field size	12 symbols	20 bytes		Variable
Field name	Preamble	Physical header	CRC checksum	User data
Comment: the actual physical size of individual components will depend on SF value				

Fig. 4.6 The physical layer frame format for DSSS systems

(real speeds in the first IEEE 802.11 specification of about 1.5 Mb/s and in Bluetooth about 1 Mb/s), eventually shared the market, the development of WLAN has followed wideband transmission, while Bluetooth has moved towards providing peripheral devices with a personal range (i.e. a few or a dozen meters), focused around the phone (smartphone) or computer (tablet, notebook).

Parameters of the OSSS radio interface and the physical frame format (Fig. 4.6):

1. Channel width: from 8 to 500 kHz, depending on the spreading factor (SF)
2. R_c chip speed: 8 –500 kc/s
3. R_b transmission rate: 30 b/s to 50 kb/s
4. Modulation patterns: the simplest recommended, e.g. BPSK, QPSK
5. P_{min} sensitivity: higher than −135 dBm
6. Centre frequency accuracy: one fourth of the chip speed, which translates, for example, to 35 ppm for the centre frequency of 868 MHz and $R_c = 125$ kc/s

With regard to the technique of implementing multiaccess control (MAC), the frame on this layer was based on the IEEE 802.15.4 recommendation [10] (also used in the ZigBee system mentioned in Sect. 1.1) in accordance with Fig. 4.7. In order to reconcile the requirement of energy efficiency and communication efficiency between terminal devices (LEP) and base station (LAP), it is recommended to use at least two of the following classes (methods) of LEP multiple access schemes:

1. Class A in which LEP, after each transmission in the LAP direction, listens for DL transmission over one or more time slots. The end devices of this class, there-

Field size (bytes)	2	0/2	0/2/8	0/2/8	0/2	variable	4
Field name	Frame control	Frame counter	Dest. address	Source address	Frame control extension	Frame payload	Hash crypto

Bit position	0-2	3	4	5	6-7	8-9	10	11-12	13	14-15
Field name	Frame type	Security enabled	AR	ADR enabled	Dest. addressing mode	Source addressing mode	More data	Transported protocol	Frame Control Extension enabled	Frame version

Fig. 4.7 Link layer access frame format of OSSS systems

fore, use asynchronous transmission, since transmission from the base station is possible only after completion of the transmission initiated by the LEP.
2. Class B, in which the devices synchronize with the LAP, listening for incoming beacon signals sent every 128 s. LAP manages the allocation of time slots in the DL transmission, deciding which individual LEP stations will listen, which results in the capability of transmitting in the synchronous mode.

4.3 Ultra-Narrowband Transmission: UNB Systems

The designation ‘ultra-narrowband’ refers to the fact that radio channels used for two-way communication are evidently narrower, compared to total available bandwidth (usually in the proportion is around 1: 100) [91]. While typical narrowband radio systems use channels with a bandwidth of several dozen kHz [92], UNB signals require BW in the range of several hundred Hz, which is achieved using finite impulse response filters (FIR), characterized by high selectivity. This type of transmission implies at least two obvious benefits:

1. It can be assumed that the character of multipath fading profile will be flat because in case of UNB channels, it is not difficult to fulfil the requirement of the channel width being much smaller than the coherence bandwidth BW_{coh} given in the formula (4.13) [93]. The delay spread, which is a measure of energy dispersion in the time domain [94] due to the nonsimultaneous arrival of multipath components, has been marked using τ_{RMS} symbol (see also Sect. 7.1.2). It is a known fact that fulfilment of this condition, i.e. $BW << BW_{coh}$, guarantees that the transmitted signal will be subject to identical attenuation over its entire channel width BW [95]. This certainty, in turn, eliminates the need for channel equalization for compensating the effects of frequency-selective fadings. It is clearly presented in Table 4.1, where BW_{coh} is shown, calculated from the relation (4.13) for typical τ_{RMS} values characteristic for a wide range of propagation environments. The comparison shows that even in the most pessimistic propagation scenario (i.e. macrocells in a hilly terrain), the coherence band is 28 kHz, i.e. about

Table 4.1 List of typical values of τ_{RMS} and resultant BW_{coh} [93] widths

	Delay blur τ_{RMS} [µs]		Coherence bandwidth BW_{coh} [kHz]	
System family	Minimum	Maximum	Minimum	Maximum
Building interiors	0.01	0.05	27.566	5.513
Satellite channel, mobile	0.04	0.05	6.892	5.513
Open space	0.2		1.378	
Suburban macrocell	1		276	
Urban macrocell	1	3	276	92
Macrocell in a hilly terrain	3	10	92	28

three orders of magnitude higher than the typical channel widths in UNB systems.
2. There is a possibility of simultaneous, interference-free transmission from many terminals simultaneously due to the availability of a large number of frequency channels. SigFox, for example, operating in a range of a total width of 192 kHz, has 1920 individual channels with an individual width of 100 Hz in this band. This fact, in turn, implies the possibility of significantly simplifying the multi-access protocol and, consequently, reducing the associated overhead in the form of an additional delay, reducing the transmission speed, eliminating energy losses for handling CCA procedures or minimizing the necessary signalling overhead.

$$BW_{\text{coh}} = \frac{\sqrt{3}}{2\pi \cdot \tau_{\text{rms}}} \tag{4.13}$$

Extremely small radio channel widths (e.g. 100 Hz in SigFox) prove classical techniques of multiaccess (e.g. FDMA, TDMA) or channel multiplexing (e.g. FDM, TDM) inefficient, since these require precise separation of individual channels in the time domain. The key parameter of these techniques is the oscillator precision, whose possible defect might generate an undesirable shift between the intended and the actual frequency. Such small channel widths in UNB systems would, therefore, imply the necessity of using very high-precision oscillators. However, this requirement would be in conflict with the premise of low cost and mass production of UNB equipment. The general opinion is, therefore, that the classical methods of multiple access or multiplexing are inadequate for the specifics of UNB systems where (see [96]) the frequency uncertainty is higher than the bandwidth itself. With inferior class of oscillators, it is therefore not possible to ensure nonoverlapping channels, as is required in traditional narrowband systems, which has led to the development of a completely new multiple access paradigm for the use of ultra-wideband UNB systems, namely, multiaccess with Random Frequency and Time Division Multiple Access (RFTDMA) described in Sect. 4.3.3.

4.3.1 UNB Systems Uplink (UL) Radio Interface

In Europe, the frequency band for the uplink UNB transmission covers the following range: 868.00–868.60 MHz ('M' band in [80], acc. to [81]; see Sect. 2.4), with a maximum ERP at 25 mW and maximum duty cycle (DC) equal to 1% for the UL and 10% for the DL. The UNB systems, however, do not utilize LBT or AFA techniques (discussed in Sects. 2.5.2 and 2.5.3). The main UNB transmission parameters include [77, 78]:

- *BW*: channel width: 100–250 Hz (600 Hz in the USA).
- Transmission rate: 100 baud (600 baud in the USA).

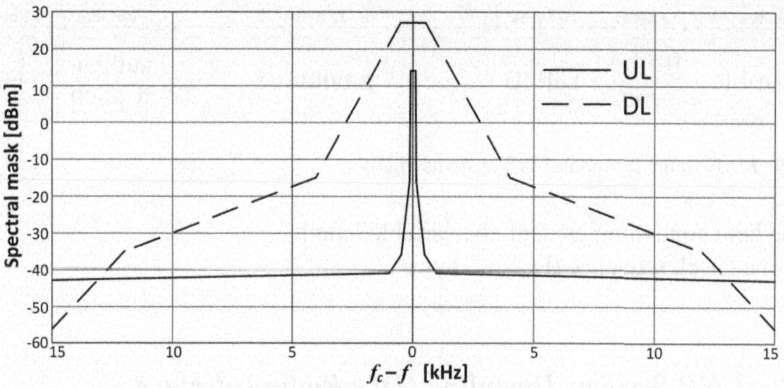

Fig. 4.8 UL and DL spectral masks for UNB systems [77]

- Modulation: BPSK.
- $P_{\min}$ sensitivity:

  −136 dBm for $BW = 250$ Hz
  −140 dBm for $BW = 100$ Hz

- Accuracy of the channel centre frequency: not applicable, greater than the channel width.
- Typical frame transmission time: 2 s (range: 1–4 s).
- A standardized channel spectral mask for UNB systems is shown in Fig. 4.8 based on the guidelines following [77] outlined in Table 4.2.

The recommended MAC frame format for UL is also defined, shown in Fig. 4.9, where individual fields are as following:

- Preamble: 4 bytes
- Synchronization field: 2 bytes
- Terminal device ID (LEP): 4 bytes
- Data field (user data): 0–12 bytes

Table 4.2 UNB systems ERP emission levels in UL and DL [77]

UL		DL	
f_c–f [kHz]	ERP [dBm]	f_c–f [kHz]	ERP [dBm]
±0.125	14	±500	27
±0.15	−16	±4	−15
±0.5	−36	±12	−35
±1	−41	±1000	−56
±15	−43	±2000	−56
±100	−46	±10,000	−56
±2000	−46		
±10,000	−46		
For RBW = 250 Hz		For RBW = 1 kHz	

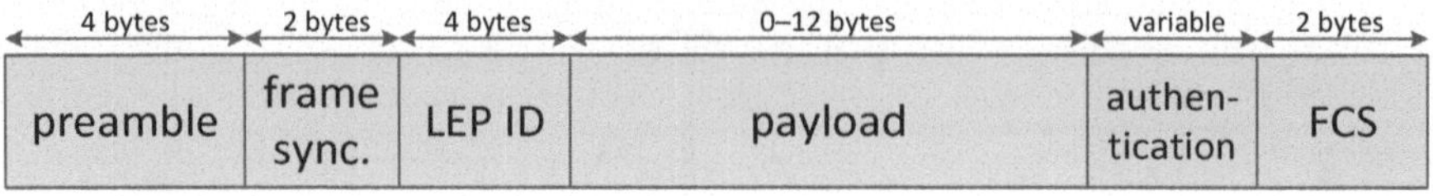

Fig. 4.9 MAC frame format in UNB systems in UL

- The field containing credentials: variable length
- Frame check sequence (FCS): 2 bytes

4.3.2 UNB Systems Downlink (DL) Radio Interface

In Europe, the frequency band for UNB 'downward' transmission covers the following range: 869.40–869.65 MHz (band 'P' as in [80, 81]; see Sect. 2.4), enabling ERP the whole 500 mW and a maximum duty cycle (DC) of 10%. The increased DC results, of course, from the fact that the base station (BS) traffic is heavily loaded due to handling a large number of terminals (UE). The main UNB transmission parameters include [77, 78]:

- A channel mask as in Fig. 4.8 based on the guidelines defined in [77] outlined in Table 4.2
- Bandwidth BW: 600–1500 Hz
- Transmission rate: 600 baud
- Modulation: GFSK
- P_{min} sensitivity:

  −124 dBm for $BW = 1500$ Hz
  −126 dBm for $BW = 1000$ Hz

  The 'downlink' MAC frame format is shown in Fig. 4.10. Individual fields mean:

- Preamble: 32 bits
- Field for synchronization: 13 bits
- Flags: 2 bits
- Frame check sequence FCS 8 bits
- The field containing credentials: variable length
- Error codes: variable length
- Data field (user data): variable length

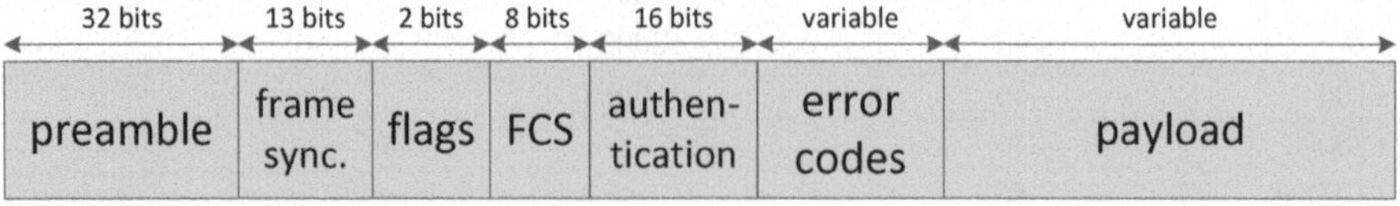

Fig. 4.10 MAC frame format in UNB systems in DL

4.3.3 RFTDMA Multiple Access Mechanism in UNB Systems

A typical method of implementing multiaccess in ultra-wideband systems is by using the Random Frequency and Time Division Multiple Access (RFTDMA). The most important features of this method are asynchronous node access (end device) to the medium, randomness in both time and frequency domain and lack of competition protocol.

Randomness in the time domain: nodes send messages at a convenient time, omitting any scanning procedure to discover spectrum parts currently being occupied. This corresponds to the Aloha technique characterized by rather low efficiency (max. 18%) typical for this method, measured by successfully completed transmissions, without collisions with transmissions from other devices. The situation in IoT applications, nevertheless, is improved by the sporadic character of transmission resulting, on the one hand, from the nature of applications that UNB systems are dedicated to (mainly based on transmissions from rarely reporting sensors) and, on the other, from the limitation of the maximum number of packets transmitted by UE per day, often imposed by the systems themselves (from several to several dozen messages per day).

Randomness in the frequency domain: unlike the classic Aloha technique, carrier frequencies are selected here from a continuous range of *BW* instead of being picked from a discrete set of channels. What it means is that the channels may also overlap partially, as discussed in detailed interference analysis, e.g. in [91]. Consequently, in order to effectively carry out the detection and demodulation process on the receiver side, effective algorithms are required for simultaneous analysis of the entire available band, with high resolution (remembering that the UL channel width in UNB systems is on the order of 100–200 Hz). Detection of activity in specific parts (channels) of the monitored spectrum takes place in an adaptive detector, located after the FFT block, which detects the spectral signatures of transmitted UNB signals. Thus, any uncontrollable frequency drifts resulting from the imperfection of the oscillator in the transmitter are not an issue, as long as the drift remains constant during the broadcasting period. For each detected transmission, the specific frequency range is then isolated by filtration and demodulated using a BPSK demodulator.

Asynchronous access, in turn, allows to skip the overhead associated with maintaining synchronization, although it leads to extremely varied levels of interference during the broadcast of a given packet, as transmissions begin and end in a completely desultory manner.

The lack of competition protocols (as a general feature of MAC protocols in both UNB and LPWAN networks) means that each UE decides on the carrier frequency of its choice, with the omission of the stage of prior monitoring/analysis of spectrum occupancy in the cell (which is often the first step in the protocols CSMA/CA). Naturally, it becomes an obvious source of possible interference when transmissions from at least two devices overlap in time and frequency. However, in order to reduce the effect of such an event, the same packet is transmitted in UNB systems several times, each time on a randomly selected frequency.

Taking into consideration the above features, the advantages of RFTDMA are as follows:

- Energy saving resulting from the absence of need for monitoring the medium (in other words, for carrying out the CCA procedure) by end devices. This feature stems from very large cell sizes, which makes the channel observation results made by a UE in a particular location not necessarily representative for UEs located in other points of the area covered by a given LAP.
- Energy saving resulting from the lack of beacon signals necessary to maintain synchronization in the network.
- No rigorous requirements as to precision of the oscillator. Due to the fact that any frequency can be selected from the serviced range, it is even possible to use the cheapest oscillators without any degradation in UE (and the whole network) functioning.

The abovementioned advantages, however, mean that despite the ultra-narrow frequency channels, there is still a risk of interference or collision between active users, as discussed in [87].

From the receiver standpoint, the monitored BW spectral range is filled with a time-frequency combination of narrowband signals randomly scattered in both of these domains. Their demodulation is based on efficient algorithms dedicated to scanning the whole band in real time, detecting the transmitted signals and interpreting them. This is accomplished with the help of an FFT block and an adaptive detector, designed to identify the spectral signatures of UNB signals. Thus, any uncontrolled frequency shifts in the transmitters are not problematic, as long as a shift remains unchanged during the packet transmission duration. For each detected transmission, its spectrum is extracted and then demodulated in a standard BPSK demodulator.

In order to increase the reliability of such a multiple access mechanism, it is assumed that each message will be transmitted up to three times, each time on a different frequency. In response, the base station (BS) transmits a reverse packet (if so provided by the communication protocol) to the device (UE) on the same frequency on which it received transmission from the uplink direction, which creates an additional simplification of the reception algorithm on the UE side in that the device is not required to analyse the whole band.

Chapter 5
The Internet of Things Narrow-Band LPWAN/UNB Systems

As frequently mentioned already, the concept of the Internet of Things has led to the development of a number of new systems and protocols characterized by a specific set of requirements that neither current cellular systems nor long distance (e.g. WiMAX) nor short term (e.g. Bluetooth, IEEE 802.11g/a/n/ac) were able to meet in full scope. These distinguishing features include, above all, the large coverage achievable through the use of various signal processing methods, resulting in increased processing gain, adaptive transmission rates (from a few hundred b/s to a dozen or so kb/s), radiated power, modulation, duty cycles, etc., so that, at the same time, the connected devices energy consumption could be minimized.

5.1 LoRaWAN (LoRa): Architecture, Radio Interface and Device Classes

'LoRaWAN', an abbreviation standing for long-range WAN, is a long-range open telecommunications system currently supported by the LoRa Alliance. Under the name LoRa, one should understand a closed, proprietary solution for the physical layer of the LoRaWAN (long-range wide area network) system [99, 100], developed by the French company Cycleo (later acquired by Semtech), based on a variation of transmission with a spread spectrum, using a multi-valued CSS method. This means that in contrast to the standard CSS technique described in Sect. 4.1.2, where a single chirp symbol encoded only one bit, the version of this method used in LoRa allows to encode from 7 to 12 bits in a symbol. In principle, the LoR is characterized by a long unattended battery life (about 10 years in favourable conditions), low transmission speeds (up to 22 kbps for SF = 7 and BW = 500 kHz and 50 kbps in the physical layer using FSK modulation) and large ranges (2–5 km in urbanized areas up to 15 km in suburban areas). The system can work, depending on the region of the world, in variously defined ISM bands, in which different rules apply, for

K. Staniec, *Radio Interfaces in the Internet of Things Systems*,
https://doi.org/10.1007/978-3-030-44846-2_5

example, regarding the effective isotropic radiated power (EIRP), duty cycle (DC), an obligatory number and width of channels, the use (or not) of LBT, AFA mechanisms or preamble length, etc. The regional specification of the LoRaWAN system [102] determines the values of these parameters individually for the following bands used in different parts of the world:

- EU 433 MHz ISM Band
- CN 470-510 MHz Band
- China 779-787 MHz ISM Band
- EU 863–870 MHz ISM Band in sub-bands (according to Table 2.7):

  M: 868–868.60 MHz (600 kHz in total), ERP = 25 mW, DC $\leq$ 1%
  P: 869.40–869.65 MHz (250 kHz in total), ERP = 500 mW, DC $\leq$ 10%

- US 902-928 MHz ISM Band
- Australia 915-928 MHz ISM Band
- AS923 MHz ISM Band
- South Korea 920-923 MHz ISM Band

5.1.1 LoRaWAN: System Architecture

LoRaWAN, in addition to the system name, is also the name of its link (medium) access layer and defines the communication mechanism of end devices (Fig. 5.1) with LoRa gateways in the star topology, using the LoRa radio interface. LoRaWAN system architecture consists of the following elements:

- End devices responsible for collecting data from sensors and meters and then – using the LoRa radio interface – passing them on to the gateways.
- Gateways are intermediary devices used for forwarding packets coming from the end devices to a network server through an IP-based backbone, allowing the transfer of large aggregate traffic, using such techniques as Ethernet or cellular networks. Several gateways can operate in a given LoRaWAN network, with the same packets being received by different gateways to increase the reliability of data delivery to the server.
- A network server, responsible for removing duplicate packets, decoding them and generating packets to be sent back to end devices.

Unlike traditional cellular networks, end devices are not linked to any specific gateway through which they could access the network. The gateways are used rather as relays of the link layer, to forward packets between devices and the network server, after adding information about the reception quality. Thus, the end device is logically connected to the network server, responsible for the detection of duplicate packets, selection of the appropriate gateway for the relaying of responses and the generation thereof. From a logical point of view, the gateways themselves are transparent to end devices.

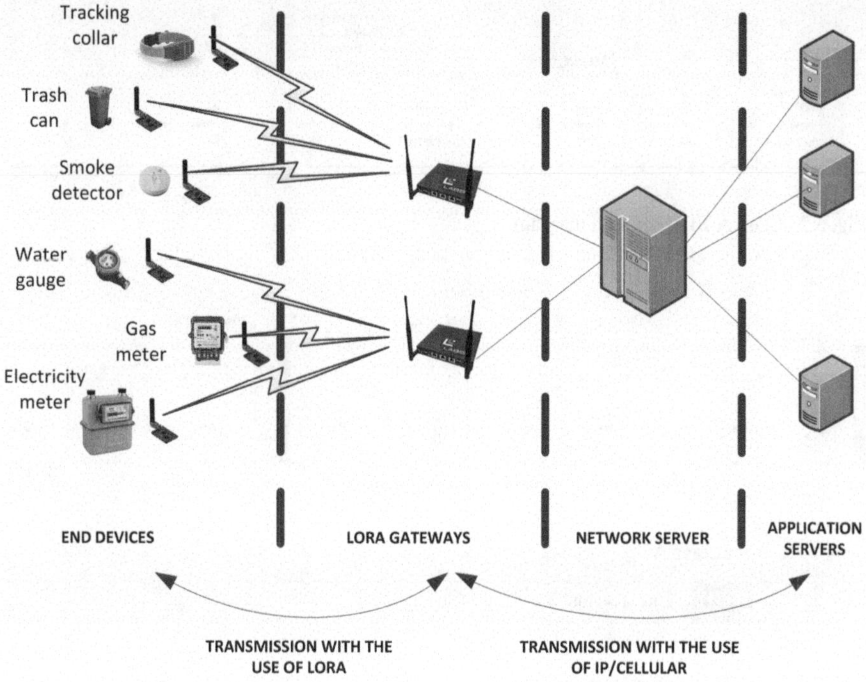

Fig. 5.1 LoRaWAN system topology

5.1.2 LoRaWAN: Device Classes

LoRaWAN defines three classes of end devices (ED), implemented according to the applications for which the ED data was intended. These classes – A, B and C – differ substantially in their work cycle which, in turn, affects their availability (from the point of view of the LoRa gateway) and energy efficiency.

Class A devices – mandatory in all ED's – transmit to the gate only when they have data to send, e.g. after the sensor reading that should be sent. After the transmission is completed, the gateway, which is always in the listening state, gets the opportunity of sending data to the ED twice, after the 'RECEIVE_DELAY1' period and then after the 'RECEIVE_DELAY2' period in the reception windows: 'RX1' and 'RX2', respectively. The end device listens in the second receiving window; however, only when in the first window it has not received any data from the gateway. It then transits to the sleep mode and waits for the data to appear in the transmission buffer from the sensors. Class A, obligatory for all ED in the LoRa system, is at the same time the most energy-efficient but also the least flexible in terms of free communication in both directions (ED ↔ LoRa gateway) due to lack of control from the gateway over the moment of starting its own transmission; the connection initiation is solely the responsibility of the end device, as shown in Fig. 5.2.

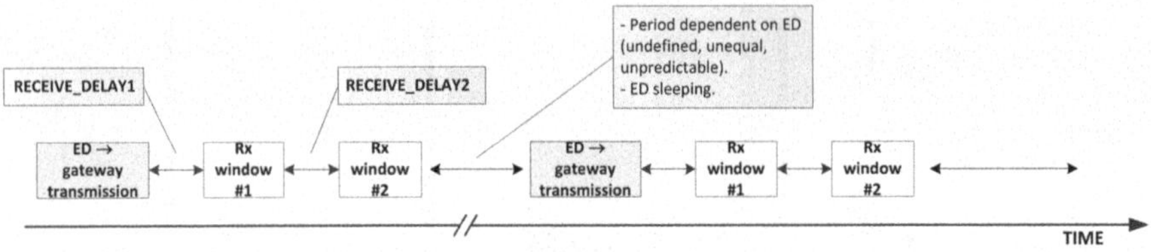

Fig. 5.2 Class A ED operation diagram

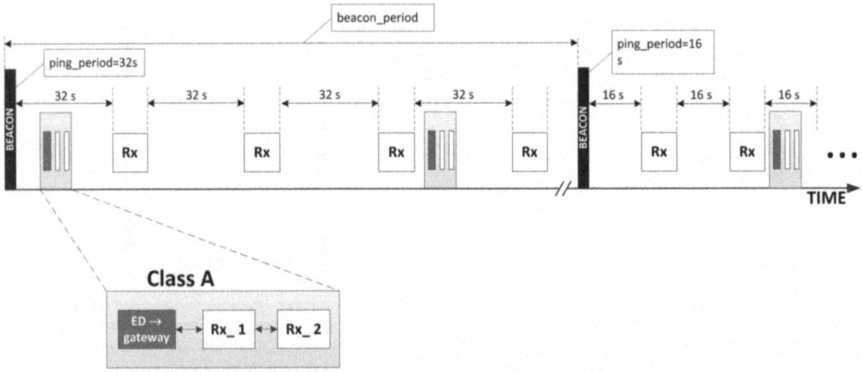

Fig. 5.3 Class B ED operation diagram

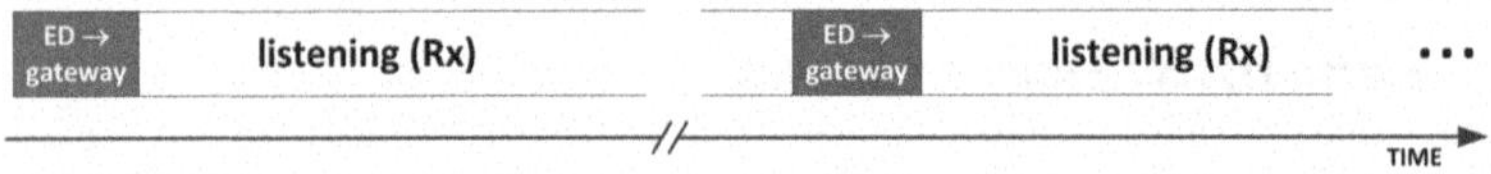

Fig. 5.4 Class C ED operation diagram

Class B of terminal equipment – optional – assumes that, in addition to the implementation of class A (also included in class B), ED will be opening then receiving windows at fixed intervals ('beacon_interval'), previously determined at the system configuration stage, in order to obtain synchronization with beacons emitted by the LoRa gateway, e.g. every 128 s (as in Fig. 5.3). The most important message carried in the beacons is the information about the listening period (shorter than the 'beacon_interval') which sets the ED's clocks to regular monitoring of transmissions from the LoRa gateway. Such synchronized end devices, parallel to the operation according to the Class A diagram, will regularly switch to the 'beacon_interval' listening. This method, due to the increased flexibility of communication between the LoRa gateway and the terminal device, is suitable for applications tolerating certain delays in the reception of messages from end devices, but which require their regular arrival of messages and control over the ED transmission cycle.

Devices belonging to the optional Class C never go into low-energy sleep mode, switching only between the states of transmission and listening (Fig. 5.4). This feature makes it, admittedly, the least energy-efficient but at the same time the most effective one in the context of the LoRa gateway ease of reaching end devices. A typical use scenario for devices of this class has a small tolerance for delays and requires immediate communication with an ED, on demand.

5.1.3 LoRa: Frame Structure, the Most Important Commands

Although LoRa modulation can be used to transmit any frame, the specification [100] determines the physical frame format within which the bandwidth and the spreading factor are held constant. The frame begins the preamble consisting of a series of upchirps, covering the entire width of the BW. In the European band, EU 863–870 MHz ISM Band, it consists of 8 symbols (for Lora modulation) or 5 bytes (for GFSK mode) [102]. The two final signals denote the sync word being a 1-byte value for distinguishing LoR networks using the same frequency ranges. The device, configured for a given word sync word, will stop listening for a particular transmission if the decoded word is different from its own. After the syncword, there are 2.25 downchirps. The appearance of the preamble transmission is depicted in Fig. 5.5 [100, 101]. After the preamble there is a PHDR header with a minimum length of 13 bytes, and a maximum of 28 bytes, optional for the DL, and obligatory for the UL. Its occurrence is called the explicit mode whereas its omission the implicit mode. The latter mode, leading to the reduction of the signalling data in the frame, is selected in situations when the transmission header is not necessary because, e.g. *MACPayload* in Fig. 5.6, coding efficiency *R*, spreading factor *SF* and information about the presence (or absence) of the *PHYPayloadCRC* field are known to both sides of the link (i.e. LEP and LAP) and unchangeable. In the explicit mode, however, regarding UL, there is both a PHDR and a checksum, computed based on the *PHYPayloadCRC* data field, whereas regarding DL, *PHYPayloadCRC* is never included.

The header also has its own cyclic redundancy check (PHDR_CRC) which enables the receiver to detect and reject packets with corrupt headers. The maximum length of the *MACPayload* field of use depends on the region and has been

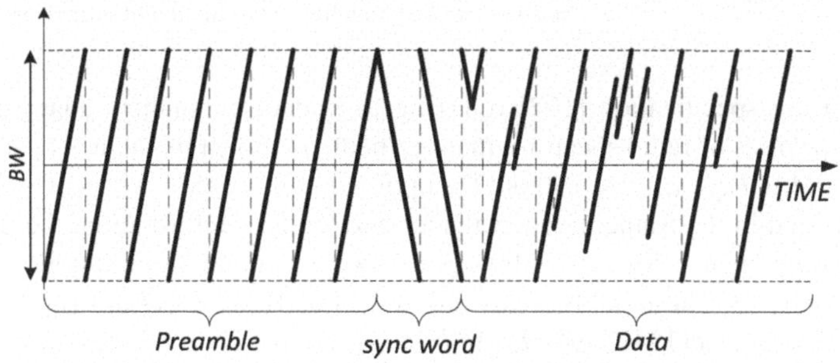

Fig. 5.5 Transmission of the preamble and data in the time and frequency domains

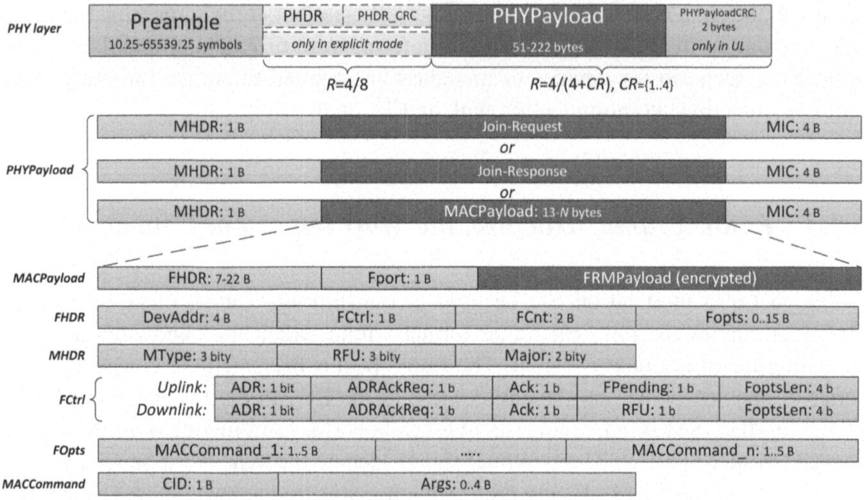

Fig. 5.6 Elements of the physical layer (PHY) frame and link access (MAC) for LoRa

Table 5.1 Maximum MACPayload sizes for EU863-870 bands [100, 102]

Transmission rate DR *(data rate)*	*M*	*N*
0	59	51
1	59	51
2	59	51
3	123	115
4	230 (250[a])	222 (242[a])
5	230 (250[a])	222 (242[a])
6	230 (250[a])	222 (242[a])
7	230 (250[a])	222 (242[a])
8:15	Undefined	

[a]Used with an assumption that the end device will never communicate with the base station via a repeater. Smaller values of M and N in front of the parenthesis are used when it is probable that ED may work with a relay; these values take into account the margin resulting from the possible protocol overhead associated with the encapsulation of packets in the relay

defined in the specification [102] referring to system parameters adapted to the regional specifics of radio spectrum management, as shown in Table 5.1, where the maximum *MACPayload* size without the optional *FOpts* field is *N* and *M* when this field is present in the frame. Even in the absence of user information, the presence of other component fields causes frames to always contain a minimum number of bytes, namely, 13. These fields are as follows: MHDR [1], DevAddr [4], FCtrl [1], FCnt [2], Tport [1] and MIC [4]. *DevAddr* in Fig. 5.6 is a short device address, FPort a field indicating the nature of the data (*FRMPayload*), so its zero value means that this field contains only MAC commands, and its non-zero value indicates that MAC commands are transported in the *FOpts* field. In this case, the value of the *FOptsLen* field is also zero. *FCnt* is a frame counter. The MIC field, on the other hand, is the message integrity code, calculated on the basis of the MHDR, FHDR, FPort fields and the encrypted FRMPayload. *MType* specifies the message type (one of eight), indicating, for example, the message direction (i.e. UL/DL), specifying whether it requires acknowledgement or whether it is a query or acceptance to join the LoRa gateway. *Major* carries information about the LoRaWAN version (currently, the only correct value is zero). *ADR* and *ADRAckReq* fields control the mechanism of the baud rate adaptation by the network server. *ACK* confirms the last received frame. The *FPending* bit, set by the gateway, indicates that there is data in the transmit buffer intended for the end device, requesting opening another receiving window by sending another UL message. *FOptsLen* determines the length of the *FOpts* field in bytes *FOpts* used to transport MAC commands with a maximum total length of 15 octets.

Command identifier (*CID*) identifies a command type. Due to a wide range of radio parameters, such as bandwidth (*BW*), maximum EIRP, duty cycle (*DC*) or occupancy time, the one method of coercing the end device (*ED*) to set these parameters in accordance with regional regulations is by the use of these commands, by means of which the gateway 'tunes' the associated end devices (*ED*). Some of the most vital commands include:

- *CID = 0x02 (LinkCheckReq)*, sent by the end device *ED*, queries the gateway for confirmation of being connected to it. The indicator with the same value (LinkCheckAns), sent back by the gateway as a response to the ED query, contains information about the power level of the signal received from the ED that allows it to assess the link quality.
- *CID = 0x03*, if sent by the gateway *(LinkADRReq)*, means a request to change the baud rate, transmit power, number of repetitions or channel. When sent by an ED (LinkADRAns) means that these changes have been accepted.
- CID = 0x04 identifies the *DutyCycleReq* command specifying the current DC. The terminal device accepts these settings by sending back *DutyCycleAns* command with the same CID.
- CID = 0x06 means *DevStatusReq* command: the gateway asks ED for the battery status and reception quality. ED responds with the *DevStatusAns* command, in which the reception quality determines the so-called margin of demodulation, i.e. the distance between the SNR of the last received message and the minimum SNR needed to receive it properly. In other words, a demodulation margin of 0 dB indicates a reception at the verge of correctness; its higher values, in turn, indicate a certain energy reserve, which increases the assessment of the quality of reception from the ED's point of view.
- CID = 0x09 (*TxParamSetupReq* command) sent by the gateway carries information about the new maximum dwell time and the maximum EIRP allowed by regional regulations, as in Table 5.2. Occupancy time means the maximum period for which ED can transmit continuously (also subject to legal regulations). The end device responds with the same CID indicator, in this case denoting acceptance of the settings (*TxParamSetupAns* command).
- Keep-alive confirmation sent by the terminal device to the gateway.

Table 5.2 Maximum EIRP values established for LoRaWAN system devices [100]

Coded value	0	1	2	3	4	5	6	7	8	9	10	11	12	13	14	15
Max EIRP [dBm]	8	10	12	13	14	16	18	20	21	24	26	27	29	30	33	36

The contents of the *FRMPayload* field, before calculating the MIC, must be encrypted. The encryption method is based on the algorithm described in IEEE 802.15.4-2006 specification, Annex B [10] using AES, with a 128-bit key length.

5.1.4 Activation of the End Device

For the end device ED to be able to work in the LoRaWAN network, it must be activated. LoRaWAN defines two ways of activating: over-the-air radio activation (OTAA) and activation by personalization (ABP). As a result of activating the ED, it should obtain the following information:

- Its address (*DevAddr*): a 32-bit identifier of the terminal device, of which 7 bits is used to identify the network, while the remaining 25 bits are the ED network address
- The application identifier (*AppEUI*): the global identifier specified in the EUI64 address space, defining in a unique manner the owner of the terminal device
- The network session key (*NwkSKey*): keys used by the network server and ED to calculate and verify the message integrity code (*MIC*) in order to ensure data integrity
- The application session key (*AppSKey*): the key used by the network server and the end device for encrypting and decrypting the data field in information messages

In the case of OTAA, the connection procedure requires the exchange of 'join-request' and 'join-accept' messages for each new session. On the basis of the 'join-accept' message, the terminal devices then retrieve the session keys (*NwkSkey* and *AppSKey*). During the procedure, these messages, as shown in Fig. 5.6, replace the 'MACPayload' field.

In the case of ABP activation, in turn, both session keys are loaded directly to the end devices.

5.1.5 LoRa Physical Layer

The modulation used in the LoRa system (also known as 'LoR modulation') is an enhanced, patented ([103]) form of chirp spread spectrum (CSS) described in Sect. 4.1.2. It turns out that, compared to other transmission techniques (e.g. FHSS or DSSS), LoRa is very useful in long-distance IoT systems applications due to its high processing gain, allowing for long-distance work or, in urban environments, over smaller distances (several kilometres) but with effective building penetration.

5.1.5.1 Modulation and Demodulation

In order to be able to encode to $SF = 12$ bits in a single LoRa chirp signal, it is necessary to define a unique frequency trajectory within a single chirp signal for each of the 2^{SF} samples, each lasting $T = BW^{-1}$ [s] and the entire signal duration of $T_s = T{\cdot}2^{SF}$ [s]. The modulation effect is achieved by assigning to each symbol a unique shift (offset) of k chips (samples) with respect to the initial frequency f_0, thus encoding the symbol value, or a given modulation state, to that shift. In this way, in order to make space for an m-state modulation, the whole bandwidth $BW = f_F – f_0$ is divided into 2^m equal intervals, each of width $\Delta f = BW{\cdot}2^{-m}$, localized at discrete frequencies: $\{f_0 + 0{\cdot}\Delta f, f_0 + 1{\cdot}\Delta f, f_0 + 2{\cdot}\Delta f, \ldots, f_0 + (2^m – 1){\cdot}\Delta f\}$. Each of these becomes a starting frequency uniquely associated to one of 2^m binary sequences. A given m-bit long k-th sequence is encoded by initializing (at time T_0) the signal upchirp at a frequency $f_0 + k{\cdot}\Delta f$, letting it rise (either linearly or exponentially) up to f_F and then abruptly reducing the instantaneous frequency down to f_0 and continuing its incrementation until the end of the chirp that occurs at time T_s.

This method helps obtain multistate modulation (in the range from $2^{SF=7} \div 2^{SF=12}$ states) of the chirp signal, as opposed to the classical, binary form of the chirp modulation (as described in Sect. 4.1.2). Thus, each encoded chirp is a cyclic version of the reference signal, shifted by k chips, as a result of which, the modulated chirp is characterized by a step change in frequency at a position of the k-th chip (sample) of its duration, as shown in Fig. 5.7a for an exemplary shift of $k = 32$ i $SF = 8$.

The chirp $\omega_k^{(SF)}$, modified in relation to its original form given by a formula (24), with the instantaneous frequency being a quadratic function of time (Fig. 5.7a), is now represented by the formula (5.1) [87],where k is the number of shifted chips (offset), while the original chirp signal is cyclically shifted by k chips/samples (Fig. 5.7b).

On the receiving side, the received signal ω_k is multiplied by an unshifted chirp signal by $\omega_{k=0}$. Demodulation of the preamble and synchronization word ‘SyncWord’, allowing to obtain the time reference necessary for demodulation of the further part of the signal is possible due to the orthogonality feature of the LoRa base signals with the same spreading factor SF. The feature states that the cross-correlation between 2^{SF} possible LoRa base signals with different offsets (m,k) is given by the relationship (5.2). The resultant signal $x_{dem}^{(SF)}$ (given by the formula (5.3) and plotted in Fig. 5.7c) is then subjected to the fast Fourier transform (FFT). The result of this operation is the spectrum with a maximum fringe located at a sample whose index corresponds to the value of the offset k (here equal to 32), which was sought for, as shown in Fig. 5.7. This accomplishes the demodulation process of the chirp. In order to present the complete picture, it should be noted that LoRa orthogonality is also preserved among signals with different SF values,

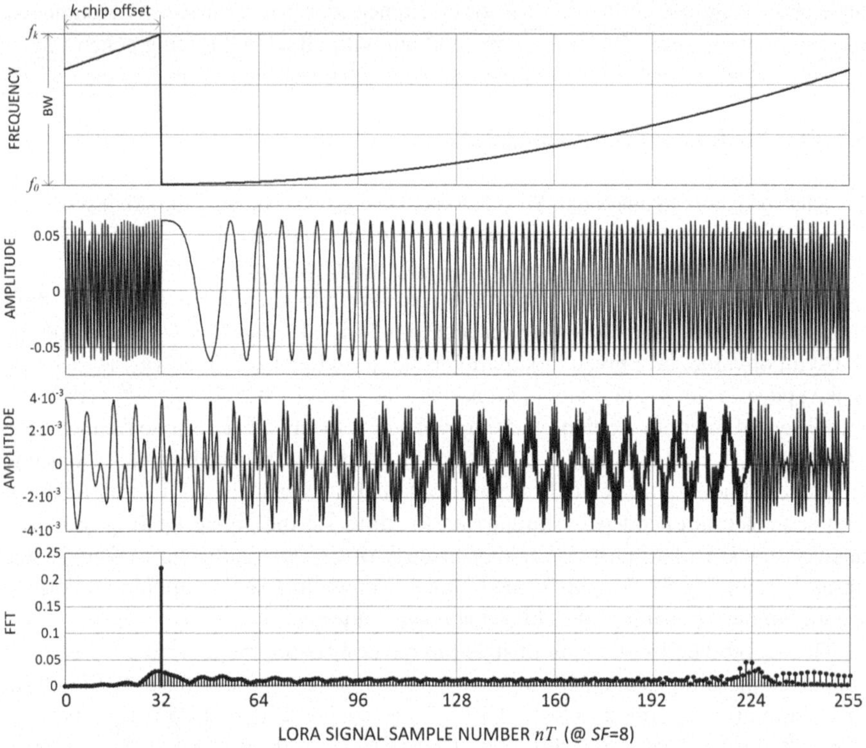

Fig. 5.7 Modulation and demodulation in LoRa: (**a**) frequency offset by k samples (chips); (**b**) a shifted chirp signal; (**c**) a demodulated signal; (**d**) FFT of the demodulated signal

regardless of their offset, presented by the relation (5.4). This feature also makes grounds for the LoRaWAN network planning process in which up to six mutually non-interfering logical networks (each assigned with its own SF value) can originate from a single gateway. This concept will be extended in Sect. 5.1.5.3.

$$x_k^{(SF)}(nT) = \sqrt{\frac{1}{2^{SF}}}\exp\left[j2\pi\cdot\frac{\left((n+k)\bmod 2^{SF}\right)^2}{2^{SF+1}}\right] \tag{5.1}$$

$$\sum_{n=0}^{2^{SF}-1} x_m^{(SF)}(nT)\cdot x_k^{*(SF')}(nT) = \begin{cases}1\; for\; m = k\\ 0\; for\; m \neq k\end{cases} \tag{5.2}$$

$$x_{dem}^{(SF)}(nT) = x_k^{(SF)}(nT)\cdot x_0^{*(SF)}(nT) \tag{5.3}$$

$$\sum_{n=0}^{2^{SF}-1} x_k^{(SF)}(nT)\cdot x_k^{*(SF')}(nT) \approx 0,\ \forall SF \neq SF' \tag{5.4}$$

In order for the described demodulation process to be carried out correctly, the following conditions must be met:

- The transmitted signal phase must be continuous, especially at the transition point. Moreover, the instantaneous phase must be identical at the beginning and at the end of the symbol, which ensures a reliable FFT result. The implementation of this requirement depends on the transmitter.
- The demodulation procedure demonstrated here is possible with ideal time-frequency synchronization, because any precision-related errors in either domain will be interpreted as an additive shift (f_{err}) relative to a correct initial frequency f_k. In the case of traditional CSS binary spatial modulation, this would not be a significant obstacle, because in order to properly distinguish logical zero from unity, one should only determine whether the frequency sweep is increasing or decreasing. In the case of the modified, multistate CSS used in the LoRa system, the information does not modulate the direction of changes in the instantaneous frequency but rather the position of the step change occurrence within the chirp signal duration, expressed by index k. For this reason a preamble consisting of a series of unmodulated chirps is sent prior to the transmission of encoded chirps, in order to achieve synchronization by means of assessing: the beginning of T_s intervals and the absolute initial frequency f_0 (as shown in Fig. 5.8). According to [103, 104], the maximum acceptable uncertainty for achieving correct synchronization is 40 ppm, which, practically, makes it possible to use lower quality, cheaper hardware solutions. After the synchronization, the decoder determines the offset (k) of the encoded symbols in relation to the reference frequency, according to the previously described procedure of multiplication by the coupled signal.

Recapping on the principle of LoRa modulation and demodulation:

- The preamble and 'SyncWord' are used to eliminate time drift and to obtain interval synchronization.
- Modulation symbols are encoded in the form of SF-bit long code words defined in the 2^{SF}-state space, in configurations determined according to the Gray scheme to ensure the greatest possible Hamming distance in order to increase the detective/corrective ability on the receiving side. This capability is controlled by selecting one of the four coding efficiencies $R = \{0.5; 0.57; 0.67; 0.8\}$ defined in the LoRa system by the formula (5.5) as a function of the Coding Rate (CR) taking values from the set $\{1, 2, 3, 4\}$.

$$R = \frac{4}{4 + CR} \tag{5.5}$$

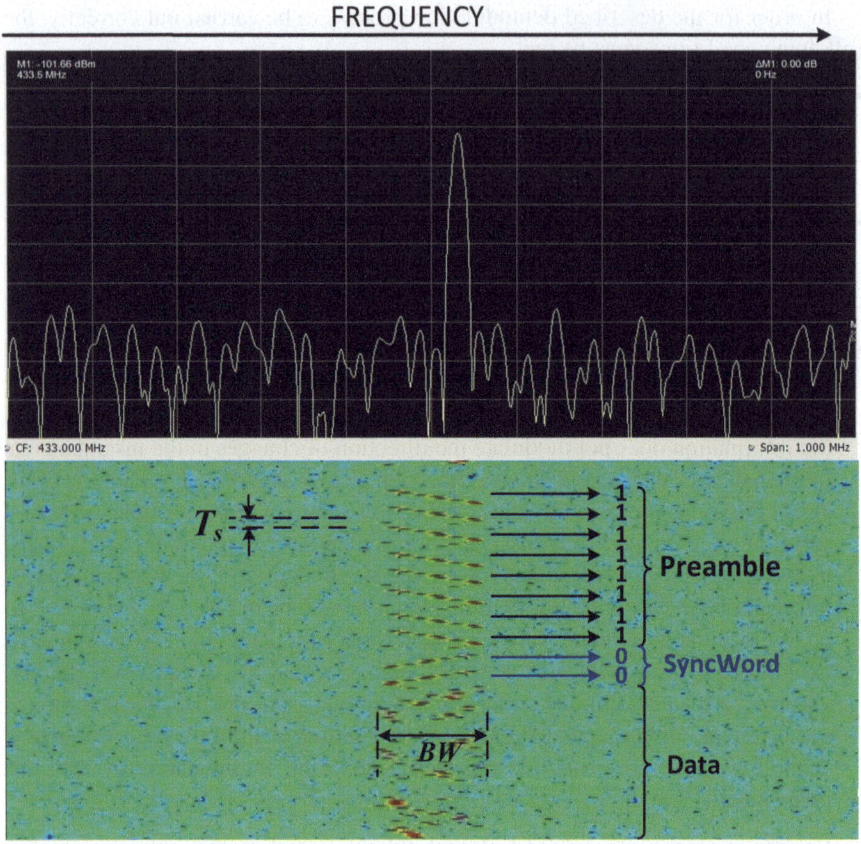

Fig. 5.8 LoRa modulation in time and frequency domains

5.1.5.2 Transmission Rate

If the spreading factor SF, the chirp symbol duration T_s and the coding efficiency R are known, one can calculate the bit rate R_b, according to the formula (5.6). The calculations, made for various combinations of BW and SF and for extreme values of R, are given in Table 5.3. The dependence determining the duration of the T_s symbol allows to determine the symbol rate R_b, as given in the relationship (5.7) and, on this basis, the resultant chip speed R_c given in formula (5.8). We may conclude that the spread data stream in LoRa is sent at the chip speed R_c [chips/s/Hz or c/s/Hz] equal to the set channel width BW. Thus a given bandwidth [Hz] (e.g. 250 kHz) corresponds to the BW chip rate [c/s] (e.g. 250 [kc/s]), i.e. in every second of the transmission, there are as many chips sent as there were Hertz transferred in that time.

Table 5.3 Transmission rates (R_b) in LoRa

BW →	*125 kHz*		*250 kHz*		*500 kHz*	
R →	*0.5*	*0.8*	*0.5*	*0.8*	*0.5*	*0.8*
SF ↓	***R_b transmission rate [kb/s]***					
7	3.42	5.47	6.84	10.94	13.67	**21.88**
8	1.95	3.13	3.91	6.25	7.81	12.50
9	1.10	1.76	2.20	3.52	4.39	7.03
10	0.61	0.98	1.22	1.95	2.44	3.91
11	0.34	0.54	0.67	1.07	1.34	2.15
12	**0.18**	0.29	0.37	0.59	0.73	1.17
Total capacity C_{b_kan} [kb/s]	**7.6**	**12.17**	**15.21**	**24.32**	**30.38**	**48.64**

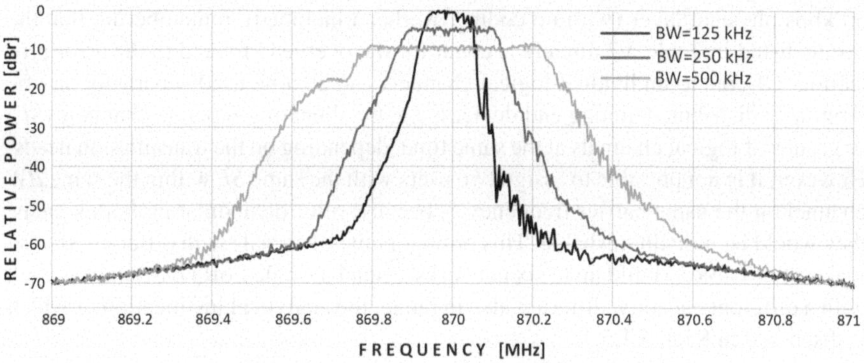

Fig. 5.9 LoRa chirp spectrum measured for different channel width [own source]

Typical bandwidths are 125 kHz, 250 kHz and 500 kHz, with the spectrum shown in Fig. 5.9. The spreading factor is selected from the range of 7–12. Due to the fact that R_c chip rate is unchanged for a given *BW*, as shown by formula (5.7), the T_s symbol/chirp duration (i.e. the time necessary to transmit 2^{SF} chips) is dependent on *SF* and grows with the increasing value of the spreading factor.

$$R_b = SF \cdot \frac{1}{T_s} \cdot R = SF \cdot \frac{BW}{2^{SF}} \cdot \frac{4}{4+CR} [\mathrm{b/s}] \tag{5.6}$$

$$T_s = \frac{2^{SF}}{BW} [s] \Rightarrow R_s = \frac{1}{T_s} = \frac{BW}{2^{SF}} [symbols/s] \tag{5.7}$$

$$R_c = R_s \cdot 2^{SF} [c/s] \Rightarrow R_c = \frac{BW}{2^{SF}} \cdot 2^{SF} = BW [c/s] \tag{5.8}$$

5.1.5.3 Aggregated Capacity, Logical Channels and Packet Transmission Length

As mentioned in Sect. 5.1.5.1, an important feature of multistate LoRa modulation is the total mutual orthogonality of chirp signals with different *SF* values. Due to this feature, individual data streams spread with different *SF* can be considered as six separate *SF*7-*SF*12 logical channels, differing in the bit rate, defined within a single physical *BW*. Thus, the total aggregated transmission capacity C_{b_agr} of a single *BW*-wide channel is the sum of the R_b bit rates obtained for each of the six available logical channels, i.e. according to formula (5.9). The total capacities for various *SF*, *BW* and *R* combinations are given in Table 5.3. The table shows that, with the complete aggregation of logic channels, LoRa modulation facilitates the net transmission capacity equal to 48.64 kb/s, exceeding that for FSK that amounts to 50 kbps physical layer (with no coding overhead included), remembering that the results listed in Table 5.3 already account for the overhead related to the error protection. Of course, individual logical channels can also be used separately and be arbitrarily distributed among end devices, e.g. by allocating separate channels (*SF*) or groups of logical channels at the same time, depending on the transmission needs. However, it is not possible to assign two users with the same *SF* within the same *BW* channel on the same carrier frequency f_c, because after demodulating both signals, they would be indistinguishable. This orthogonality of signals with different *SF* also makes it possible to build up to six networks around a single LoRa base station, each with a different capacity, differing also in range, the analytical justification of which is discussed in Sect. 5.1.5.1.

$$\begin{aligned} C_{b\_agr}(BW) &= \sum_{x=7}^{12} R_{b\_SFx}(BW) = \\ &= R_{b\_SF7} + R_{b\_SF8} + R_{b\_SF9} + R_{b\_SF10} + R_{b\_SF11} + R_{b\_SF12}\,[\mathrm{b/s}] \end{aligned} \tag{5.9}$$

For example, the total channel capacity with $BW = 250$ kHz and $R = 0.8$ is:

$$C_{b_{agr}}(BS = 250\,kHz) = 10.94 + 6.25 + 3.52 + 1.95 + 1.07 + 0.59 = 24.43\,[kb/s]$$

The selection of the system key parameters, namely, *BW*, *SF* and *R*, should result from the criteria related to the quality of the radio channel; a stronger code protection (expressed in a smaller *R* values) and higher *SF* values should be set for terminals in locations with worse propagation (usually at long distances from the base station) and/or interference conditions. In these more challenging circumstances, it is also recommended to reduce the width *BW* of the channel in order to increase the receiver's sensitivity P_{min}, due to a proportional reduction in the amount of thermal noise, thus increasing the resulting *SNR* value. In more favourable propagation and interference conditions, it is advisable to use smaller *SF* and larger *BW* and *R* values to obtain higher transmission rates, in accordance with the formula (5.6). The quantitative impact of such settings on the Packet Error Rate (*PER*) will be presented in

Sect. 5.2, based on a series of measurements made in a laboratory environment ensuring controlled interference and multipath conditions.

One of the most important criteria to be followed when selecting an appropriate *SF* value is also the duration of the T_p packet at a given bandwidth. Representation of a single bit by means of multiple chips means that these will be transmitted either at higher rates, increasing the bandwidth occupied by the signal (*BW*), or with the same *BW*, increasing the time necessary to transmit the information (T_p). Knowing the duration of a single T_s symbol (chirp), given by the formula (5.7), it is possible to calculate a single packet broadcasting time, containing *M* bytes in the data field ('PHYPayload' in Fig. 5.6). The total transmission time of the packet, therefore, consists of the transmission time of the T_{pre} preamble and *M* bytes of T_M usable data with an overhead. The preamble transmission, for all modem configurations, is given by the formula (5.10), where n_{pre} is the preamble length expressed in the number of symbols (chirps). As shown in [102], the preamble size in the three most popular bands, i.e. EU 863–870 MHz, US 902-928 MHz and EU 433 MHz, equals 8 symbols for LoRa modulation and 5 bytes for GFSK modulation (excluding the US 902-928 MHz band, where GFSK modulation does not apply). The number of symbols forming the data field of the packet and the header (N_{p_h}) and the resulting time of their transmission T_M are defined by the formulas (5.11) and (5.12) [105], respectively. Thus, the total transmission time of the T_p packet can be expressed by the relationship (5.13).

$$T_{pre} = \left(n_{pre} + 4.25\right) \cdot T_s \tag{5.10}$$

$$N_{p\_h} = 8 + \max\left(\frac{8M - 4SF + 28 + 16 - 20H}{4\left(SF - 2DE\right)}\left(4 + CR\right);0\right) \tag{5.11}$$

where

- *M* – number of bytes in the data field in bytes, given in Table 5.1
- *H* – a coefficient equal to '0' when the header is present and '1' when it is missing
- *DE* – a coefficient equal to '1' for the optimization enabled at low transmission rates and '0' without optimization

$$T_M = N_{p\_h} \cdot T_s \tag{5.12}$$

$$T_p = T_{pre} + T_M \tag{5.13}$$

5.1.5.4 Processing Gain, SNR, Sensitivity and MCL

As mentioned in Chap. 4, each of the narrowband IoT systems has a unique technological feature that allows it to obtain either a substantial processing gain G_p (e.g. LoRa, Weightless-P, Ingenu, NB-IoT) or the low noise power (e.g. in SigFox thanks to 100-hertz channels). In LoRa system, the processing gain can be calculated by comparing the following: the transmission rate R_b, calculated according to the formula (5.6) for $CR = 1$ (i.e. for the coding efficiency $R = 0.8$) and BW (125 kHz, 250 kHz, 500 kHz). The formula (22) was used for calculations, assuming rectangular pulses in the time domain for calculating the channel width before spreading BW_{inf} according to the formula (5.14) and possible bandwidths after the data spread (or BW_{SS}). The SNR ratio, on the other hand, can be obtained by referring to the SNR level of 3 dB, appropriate for the simplest two-state modulations, such as BPSK or FSK, according to the relationship (5.15). This ratio will then be used to calculate P_{min} defined by the formula (5.16). The MCL signal dynamics, which is also a measure of the maximum propagation attenuation to be experienced between the LEP end device and the LAP base station, is calculated with Eq. (1), assuming for the LEP devices an effective isotropic radiated power (EIRP) equal to 14 dBm, in accordance with the guidelines for the 'M' band, presented in Table 2.11 for EU863–870 MHz band. Values of individual parameters G_p, SNR, P_{min} and MCL were listed in Tables 5.4, 5.5, 5.6 and 5.7.

Table 5.4 The processing gain G_p [dB] in LoRa system

	The processing gain G_p [dB] $R = 0.8$					
SF →	***7***	***8***	***9***	***10***	***11***	***12***
BW ↓						
125 kHz	11	13	16	18	21	**23**
250 kHz	11	13	16	18	21	**23**
500 kHz	11	13	16	18	21	**23**

Table 5.5 Signal-to-noise SNR ratio [dB] in LoRa system

	Signal-to-noise (SNR) ratio [dB]											
SF →	***7***		***8***		***9***		***10***		***11***		***12***	
R → ***BW ↓***	***0.8***	***0.5***	***0.8***	***0.5***	***0.8***	***0.5***	***0.8***	***0.5***	***0.8***	***0.5***	***0.8***	***0.5***
125 kHz	−8	−10	−10	−12	−13	−15	−15	−17	−18	−20	−20	**−22**
250 kHz	−8	−10	−10	−12	−13	−15	−15	−17	−18	−20	−20	**−22**
500 kHz	−8	−10	−10	−12	−13	−15	−15	−17	−18	−20	−20	**−22**

Table 5.6 Sensitivity P_{min} [dBm] in LoRa system

	Sensitivity P_{min} [dBm]											
SF →	*7*		*8*		*9*		*10*		*11*		*12*	
R → *BW ↓*	*0.8*	*0.5*	*0.8*	*0.5*	*0.8*	*0.5*	*0.8*	*0.5*	*0.8*	*0.5*	*0.8*	*0.5*
125 kHz	−124	−126	−126	−128	−129	−131	−131	−133	−134	−136	−136	**−138**
250 kHz	−121	−123	−123	−125	−126	−128	−128	−130	−131	−133	−133	−135
500 kHz	−118	−120	−120	−122	−123	−125	−125	−127	−128	−130	−130	−132

Table 5.7 MCL signal dynamics [dB] in LoRa system

	MCL dynamics [dB] (=EIRP$_{=14dBm}$-P$_{min}$)											
SF →	*7*		*8*		*9*		*10*		*11*		*12*	
R → *BW ↓*	*0.8*	*0.5*	*0.8*	*0.5*	*0.8*	*0.5*	*0.8*	*0.5*	*0.8*	*0.5*	*0.8*	*0.5*
125 kHz	138	140	140	142	143	145	145	147	148	150	150	**152**
250 kHz	135	137	137	139	140	142	142	144	145	147	147	149
500 kHz	132	134	134	136	137	139	139	141	142	144	144	146

$$BW_{\text{inf}} = 2 \cdot R_b \tag{5.14}$$

$$SNR = 3dB - G_p \tag{5.15}$$

$$P_{\min} = 10\log\left(k \cdot T \cdot BW\right) + NF_{=7dB} + SNR \tag{5.16}$$

5.2 Weightless (-P): Architecture, Radio Interface and Device Classes

'Weightless (-P)' is an open communication standard dedicated for handling long-range links under conditions of strong propagation attenuation, in the presence of interference, developed and supported by Weightless SIG. The system has been designed for two-way transmission, fully synchronous, for the needs of long-range Internet of Things networks with a limited bandwidth and reduced delay requirements, adapted to work in ISM frequency bands below 1 GHz, as presented in Table 5.8 [106]. Due to the legal restrictions in force in Europe, the main focus (also in Sect. 7.2) will be placed on the 'V band' (863–870 MHz), in particular on 'M' and 'P' sub-bands according to Table 2.11, for which the following restrictions on ERP and duty cycle (*DC*) have been defined:

- Sub-band 'M': 868–868.60 MHz, ERP = 25 mW, $DC \leq 1\%$
- Sub-band 'P': 869.40–869.65 MHz, ERP = 500 mW, $DC \leq 10\%$

This system is – alongside LoRa – one of the leading 'players' in the space of LPWAN systems, as illustrated in Fig. 5.10.

Similar to the LoRa system, the Weightless (-P), recently renamed to 'Weightless', also uses the spread spectrum transmission to improve interference immunity parameters; in this case, however, the spectrum is spread directly (*DS*, see Sect. 4.1.1), using the bit interleaving techniques and data randomization (whitening). The spreading effect, however, is used only in the OQPSK modulation mode. In the second mode, GMSK, there is no spectrum spread (i.e. the dissipation factor $SF = 1$). In the former mode, two *SF* values are available: 4 (Table 5.9) and 8 (Table 5.10).

Table 5.8 Work bands envisaged for the Weightless system (-P) [106]

Band index	Frequency range [MHz]	Band name
I	138.20–138.45	168 MHz
I bis	169.40–169.60	169 MHz
II	314.00–316.00	314 MHz
III	430.00–432.00	430 MHz
III bis	433.05–434.79	433 MHz
III ter	470.00–510.00	470 MHz
IV	779.00–787.00	780 MHz
V	863.00–870.00	868 MHz
V bis	870.00–876.00	873 MHz
VI	902.00–928.00	915 MHz
VI bis	915.90–916.90	915 MHz
VI ter	920.50–929.70	923 MHz

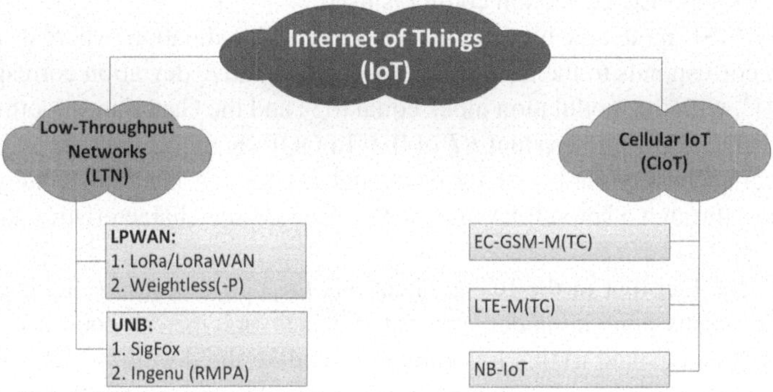

Fig. 5.10 The Internet of Things systems space

Table 5.9 Spectrum spreading sequences in the Weightless(-P) system for (4.1) mode

Input bit.	Chip sequence ($c_0 \ldots c_3$)
0	1010
1	0101

Table 5.10 Spectrum spreading sequences in the Weightless(-P) system for $(8.1)_k$ mode

k index	Input bit.	Chip sequence ($c_0 \ldots c_7$)
0	0	1011 0001
	1	0100 1110
1	0	0110 0011
	1	1001 1100

Preamble	**Synchronization word**	**PHY user data fields**
2 bytes (GMSK) 32 chips (OQPSK)	‚SyncWord' 3 bytes	0...255 bytes

Fig. 5.11 packet format in the Weightless(-P) system

For $SF = 8$, two sets of spreading sequences were defined, i.e. $(8,1)_0$ and $(8,1)_1$, used with even and odd bits, respectively. In the synchronization word, however, only $(8.1)_0$ sequence is used.

The Weightless(-P) data packet consists of the following: a preamble and a 24-bit synchronization word, followed by a PHY field containing user data, as shown in Fig. 5.11. In the GMSK modulation mode, the preamble consists of the hexadecimal string 0xAAAA, while in the OQPSK mode, the preamble consists of only zeros (occupying 1 or 2 bytes, respectively, for the spreading factor of 8 and 4). The structure of the synchronization word, in turn, is inconsistent and depends on whether pilot signals and FEC protection coding is used.

In the GMSK mode, the bits are subjected to FSK modulation, where its negative deviation corresponds to the logical '0', while its negative deviation corresponds to positive '1', with the modulation index equal to ½ and the Gaussian smoothing filter with the bandwidth-time product *BT* of 0.3. In OQPSK mode, the chip sequence is first converted into symbols (−1 for c = 0 and + 1 for c = 1) and then subjected to modulation through a smoothing filter with raised cosine characteristics and a roll-off factor of 0.8.

Due to the fact that in the IoT applications the uplink traffic (UL) is of major concern, typically containing data generated by sensors, meters, counters, etc., this chapter will be devoted to this link direction. Unlike the downlink (DL), in which all transmissions can be scheduled and synchronized between base stations, the UL requires support for transmissions generated by a large number of mutually non-synchronized end devices (ED). In order to provide larger total capacity, TDMA and FDMA multiple access schemes are utilized in uplink using narrow channels with a width of 12.5 kHz. Working in these channels, on the one hand, means a lower transmission rate but, on the other, the ability to define a larger number of logical channels, which facilitates the increase of multi-access efficiency on the given cell scale. In the UL, the base station can allocate channels either in a 12.5 kHz wide grid or in the form of aggregated eight channels yielding a channel with a total width of 100 kHz. Both channel versions, called narrowband (NB) and wideband (WB) in combination with two possible modulations (OQPSK and GMSK), result in a total of eight operational modes, corresponding to the nominal rates listed in Table 5.11. Their measured spectra, normalized to the peak value of a P_{Tx} transmitter (here equal to 14 dBm), are presented in Fig. 5.12.

The Weightless(-P) system is one of the three solutions of the Weightless SIG consortium, which has been successfully implemented on the market, with a confirmed range of approx. 2 km in favourable LOS conditions in cities.

Table 5.11 Relationships between modulation, transmission rate and channel width in the Weightless(-P) system

Channel width *BW*	Modulation	Encoding efficiency *R*	scattering factor *SF*	Bitrate R_b	Operating Mode[a] OM_x
12.5 kHz (narrowband mode, NB)	OQPSK	0.5	8	0.625 kb/s	$\boldsymbol{OM_1}$
		0.5	4	1.25 kb/s	$\boldsymbol{OM_2}$
	GMSK	0.5	1	5 kb/s	$\boldsymbol{OM_3}$
		1	1	10 kb/s	$\boldsymbol{OM_4}$
100 kHz (wideband mode, WB)	OQPSK	0.5	8	6.25 kb/s	$\boldsymbol{OM_5}$
		0.5	4	12.5 kb/s	$\boldsymbol{OM_6}$
	GMSK	0.5	1	50 kb/s	$\boldsymbol{OM_7}$
		1	1	100 kb/s	$\boldsymbol{OM_8}$

[a]According to the OM definition in **Chapter 6.2**

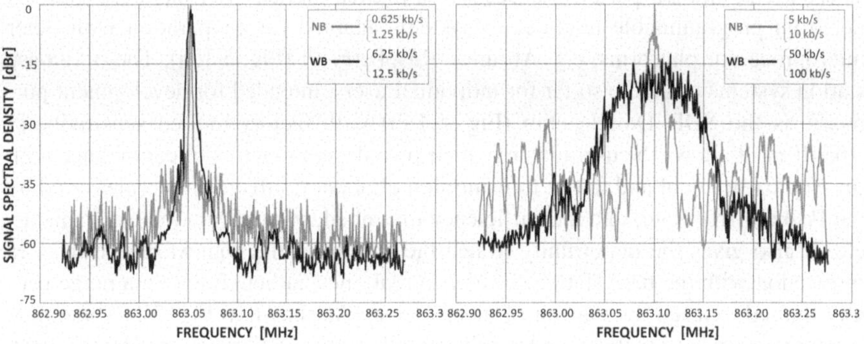

Fig. 5.12 Weightless(-P) system signal spectra measured for modes: (left) NB (12.5 kHz); (right) WB (100 kHz)

The Weightless-N standard (officially approved in 2015) assumed operation in ultra-narrowband mode in the frequency band below 1 GHz, using DBPSK modulation, but only in the UL direction. This restriction did not meet the approval of the user community interested in using the IoT systems, as described in more detail in Sect. 2.3.2. As a result, the system was eventually recalled, despite successful range tests carried out in 2015 in London. Its target range was to be 5 km in urban conditions [87].

The Weightless-W standard (officially approved in 2013) was designed to work within the 470–790 MHz digital dividend in the so-called white TV spaces. The use of different modulations was envisaged, from DBPSK to 16-QAM. The rate adaptation was arrived at via spectrum spreading at several *SF* levels (up to 1024). The disproportion in the link energy budget, resulting from EIRP being about 20 dB higher in the base station relative to UE's EIRP, was compensated by using 64 times narrower frequency channels for UL traffic compared to DL traffic, leading to the reduction in the noise level in the channel by approx. 18 dB. The target range was to be 5 km and 10 km for urban conditions, respectively, for indoor and outdoor conditions [87].

5.3 SigFox: Architecture, Radio Interface and Device Classes

The SigFox system is one of the main representatives of the class of ultra-narrowband systems (UNB, described in Sect. 4.3), where the service coverage has the peak priority, leaving out the transmission rate as a matter of lower importance. It is a product of a consortium bearing the same name [107], authorizing specific entities (companies) to which it provides base station modules and permits to use the systems as operators. End users, who are interested in using SigFox in their own networks (e.g. sensor, counter, etc. networks), become clients to these entities [108]. These users, on their part, have access to a limited market offer on client stations, i.e. from programmable modules to systems adapted for cooperation with other microprocessor platforms, e.g. Arduino, Raspberry Pi (Fig. 5.13a). The only base station systems available so far for individual users, intended for development purposes, is the SDR Dongle chip (Fig. 5.13b) with factory-limited sensitivity of −65 dBm. It allows the user to verify their own design in terms of compliance with the SigFox protocol and to test transmission efficiency, effectiveness (measured by the Packet Error Rate) and the sufficiency of embedded confirmation mechanisms, etc. It also gives the opportunity to test the system in the context of end devices interaction with the base station, similar to real ones, although within a range usually limited to a few to a dozen of meters or so. As shown in Fig. 5.14 (blue, live coverage; purple, countries under roll-out), the area of Europe has already been largely covered by the operating range of SigFox operator networks, which is by far due to long ranges obtained from a single base station.

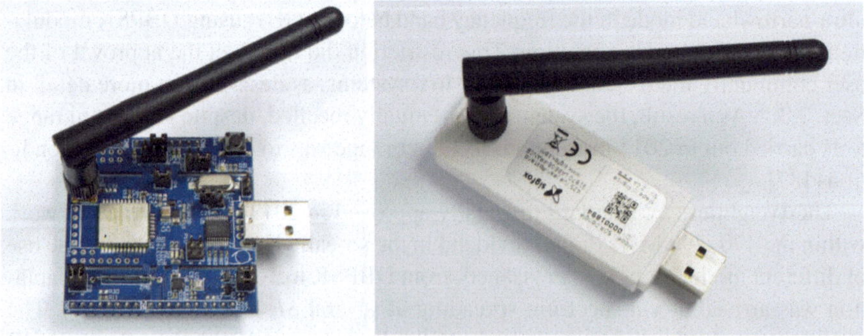

a) b)

Fig. 5.13 SigFox exemplary devices: hardware implementations: (**a**) a client station (Digi-Key); (**b**) a limited-sensitivity developer base station (Digi-Key)

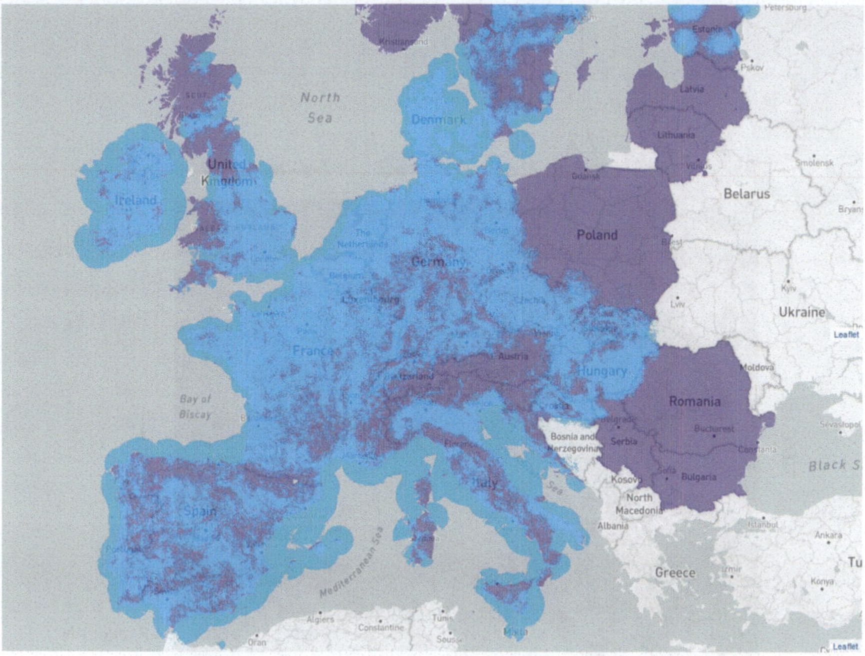

Fig. 5.14 Coverage of Europe by networks operating in the SigFox standard. Status as of 2020 [available at www.sigfox.com]

Significant energy efficiency is required from SigFox system devices, like from any other UNB system (e.g. Ingenu, Weightless-N), which is strongly emphasized in the RFTDMA multiple access protocol (described in detail in Sect. 4.3.3), in which the transmission channel is randomly selected for time and frequency without initiating any competitive mechanisms. The closest equivalent of RFTDMA is the ALOHA protocol with no pre-listening for channel occupancy. However, unlike the classic ALOHA, carrier frequencies in UNB systems are selected from a continuous range (here, with a total width of 192 kHz, as described below) instead of a predefined discrete set of channels. This feature allows for the use of weaker quality transmitters in the terminals, whose uncertainty of the generated frequency is greater than the channel bandwidth *BW* alone (which is only 100 Hz in the UL link). This feature provides another advantage of such systems, namely, reduced costs of transmitting and receiving devices, although negatively affecting the values of the *NF* (noise factor) and the *IM* (implementation margin), to the detriment of the link budget. Each data packet is repeated on the UL three times, each time on a different, random channel, whereby the transmission of each individual message takes 6.24 s. This, in turn, results in processing gain G_p resulting from these repetitions equal to approximately 4.8 dB. The measured time-spectral structure of the transmission in the SigFox system for the UL is presented in Fig. 5.15, where *PSD* stands for the

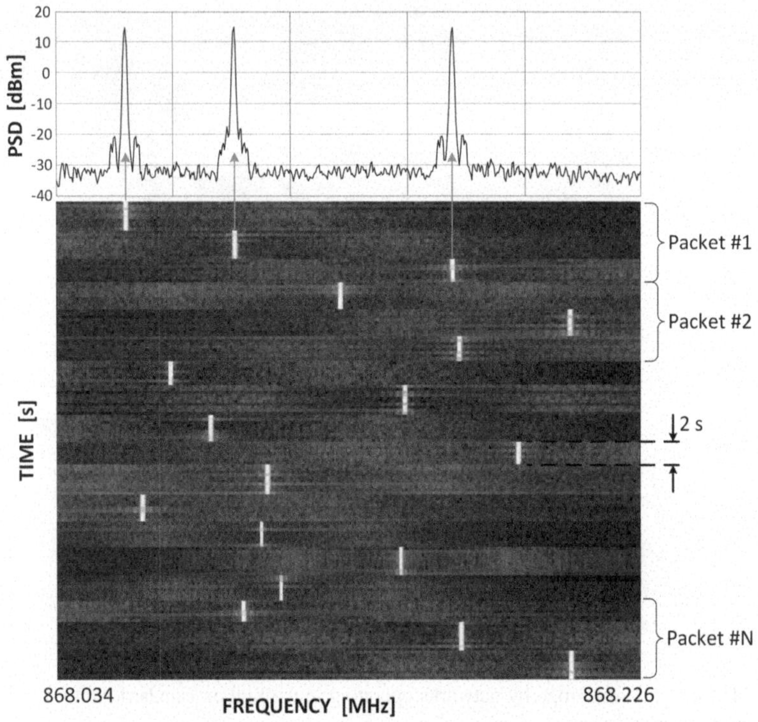

Fig. 5.15 Time-spectral structure of transmission in SigFox UL

power spectral density. It can be seen here how a single packet is carried on three consecutive frames scattered randomly in the available spectrum.

The most important features of the SigFox system are [108]:

- Operation in a fragment of the 'M' sub-band (according to Table 2.11), covering a total width of 192 kHz in the range between 868.034 MHz and 868.226 MHz.
- Nonsymmetrical two-way transmission (UL/DL).
- Bandwidth in UL transmission: 100 Hz.
- Bandwidth in DL transmission: 600 Hz.
- Network topology: star.
- The maximum permitted EIRP is 14 dBm (25 mW).
- Applied modulations: DBPSK (in the UL link), GFSK (in the DL link).
- R_b transmission rate in the UL link: 100 b/s.
- R_b transmission rate in the DL link: 600 b/s.
- P_{min} sensitivity, at $NF = 5$ dB in the DL or 3 dB in the UL, $SNR = 7$ dB and G_p processing gain of 4.8 dB, is approx. −144 dBm in the UL and −134 dBm in the DL.
- Attenuation dynamics $MCL = 158$ dB in the UL and 161 dB in the DL.

- Total length of the transmission frame: 26 B (including 12 B user data packet, payload). Detailed frame structure in the UL/DL links is presented in Sect. 4.3.
- UL frame transmission time: 2.08 s.
- Number of transmissions of a single packet: 3.
- System limitation of the maximum number of messages sent: 140 packets/day. It results from the necessity of meeting the condition $DC = 1\%$. This constraint means that, given the transmission time necessary to transmit a single message (repeated three times) of c.a. 6 s, the total transmission duration in an hour is limited to 36 s, a time just sufficient to emit statistically almost six messages. On a daily scale, this translates to a maximum of about 140 messages (packets).
- Operational range (coverage): up to 50 km in open areas, up to 5 km in built-up areas.

Chapter 6
Performance Measurements Methodology for LTN IoT Systems

6.1 Electromagnetic Interference in Internet of Things Networks

The problem of possible electromagnetic interference in IoT networks that deliver the signal in the access segment between the sensor/gauge layer and the LAP base stations has been noticed and described in ETSI TR 102 691 [52] – EMC. The document deals with the topic in general, presenting the potential goals and consequences of interference in the operation of M2M system and potential preventive measures. Two main interference groups were identified: intentional and unintentional. The purpose of intentional attacks is usually the desire to access or modify telemetric data. Non-intentional interference, on the other hand, identified with negative electromagnetic interactions, may lead to loss of communication between LEP devices and LAP base stations. Properly planned deployment of LAP stations in a given area should reduce the likelihood of such an event to a large extent. Interference from sources other than the transmitters of the tested system, sharing the same band frequency, is much more difficult to predict and control.

IoT systems operation in unlicensed ISM bands is, on the one hand, associated with the lack of license fees; on the other hand, sharing bands causes unavoidable rise of background noise as more and more new transmitters are being launched, including also wireless microphones, remote-controlled toys, etc. The problem of raising this noise was debated, among others, in [109] showing that, compared to previously performed (i.e. in 2004) wideband measurements in the 200 MHz–3000 MHz band in urban conditions, described in [110], the measurements carried out in 2016 indicate the emergence of new sources and a significant concentration of those already existing over the last decade, characterized by unpredictable time characteristics and not qualifying as additive white Gaussian noise (AWGN). Cellular systems (GSM, UMTS, LTE) and DVB-T have been identified as sources of expected transmissions. The character of some events resulting in the emission of radiation is accidental and difficult to describe using statistical models.

K. Staniec, *Radio Interfaces in the Internet of Things Systems*,
https://doi.org/10.1007/978-3-030-44846-2_6

Examples of such events are passage of a car emitting EM waves from inadequately shielded onboard electronics and control signals transmitted while remotely opening doors, gates and home automation systems. Indoors, the sources of interference are computer hardware, server rooms in particular, wireless local area network (WLAN) or peripherals communicating via Bluetooth. Increased background noise indoors can be particularly problematic for IoT systems which, by definition, are dedicated mainly to acquiring data from these locations from modems placed directly near or on a measuring device.

In the course of analogous studies published in [111], the focus was only on the ISM band 868.0–868.6 MHz ('M' band in Table 2.11) with measurements taken in the city of Aalborg in Denmark. It was observed there that 22% of the collected samples should have been qualified to a high level (i.e. above −105 dBm), among them also numerous −65 dBm power samples. The generalized extreme value distribution normally used to describe extreme phenomena, most commonly used in meteorology and hydrology, was deemed the most suitable model for such a probability density profile. Simulations of the impact of these interferences on LoRa and SigFox systems operation have proved that they cause a decrease in the effective range, respectively, down to the level of 78% and below 50%, relative to the coverage in the absence of interference. It means that even though each SigFox message is repeated three times at a different frequency, this system turns out to be more susceptible to interference than the Lora system, in which a high processing gain, in the range of 11–23 dB (shown in Table 5.4), is responsible for the system's robustness against interference.

Measurements of the interference effect on the LoRa devices, carried out for the influence of a harmonic signal and documented in [105], allow to state that this signal is not a noticeable source of interference until it exceeds the LoRa signal strength by 5 dB for $SF = 7$ and 19.5 dB for $SF = 12$, with the coding efficiency $R = 0.67$ (i.e. for $CR = 2$).

The incremental trend of implementing various LPWAN systems sharing the same frequency bands makes it necessary to develop unified procedures for their performance testing. The problem is all the more important because of the access to frequency resources guaranteed by the definition of unlicensed bands in a way that is not coordinated by any supervisory authority. The burden of implementation of collision-free multiple access, using LBT or AFA techniques discussed in Sect. 2.4, has thus been made a responsibility of systems. Nevertheless, in the case of high density of broadcasting devices, of the order of tens/hundreds of thousands per city, which is the assumption formulated for the mass Internet of Things (e.g. in [112]), simultaneous disturbances are bound to occur. From the point of view of IoT client devices, the character of these interferences will be expressed in the systematically rising background noise, leading to a gradual reduction of carrier to noise and interference (CNIR).

The following chapters will be devoted to the presentation of the author's methodology for measuring the transmission efficiency recommended for IoT narrowband systems. Chapter 7 will present the results of measurements carried out in the

Electromagnetic Compatibility Laboratory of the Wrocław University of Science and Technology (ECL, WUST), for selected LPWAN systems.

It is recommended to make performance/immunity measurements in two variants, assuming, in both cases, the Packet Error Rate as a quality measure of reception and thus immunity:

- In the presence of noise and electromagnetic interference at various power levels and constant time-spectral characteristics (in Sect. 6.2)
- Under extremely multipath propagation conditions (in Sect. 6.3)

In both cases the full measurement cycle was broken down into three stages:

- Stage 1: preparation of the measurement setup
- Stage 2: tuning of the hardware settings for measurements
- Step 3: PER measurements

6.2 Measurements of Immunity to Noise and Interference

Due to the possible effect of the gradually rising electromagnetic background, as the penetration of the telemetry market by the IoT radio systems continues, the measurement should be carried out in rooms providing the maximum possible:

- Electromagnetic isolation from external sources
- Elimination of multipath propagation effect resulting from numerous of the radio wave reflections from the walls of the room or objects located in the room

Both of the above conditions are satisfied in an anechoic chamber, whose primary and basic purpose is to study radiation patterns of various objects emitting EM waves (including mainly antennas). The anechoic chamber, a part of the WUST ECL infrastructure, guarantees 85 dB of shielding effectiveness, with the test setup proposed for immunity and performance tests presented schematically on the plan of the chamber in Fig. 6.1 and documented in photographs in Fig. 6.2.

Stage I: Preparation of the Measurement Stand

1. **Arranging the transceiver system of antennas**: in order to eliminate the influence of the device under test (DUT) which could occur as a parasitic energy radiation by onboard electronics, only antennas of the devices marked in the drawing as '$\text{antenna}_{\text{Tx}}$' and '$\text{antenna}_{\text{Rx}}$' should be located in the chamber, attached to their corresponding DUTs by means of low-loss coaxial cables of minimum length required to connect to the equipment outside the chamber during the measurements. The cables should be led outside the chamber through well-shielded feedthroughs in the chamber wall placed under a layer of absorbers. The distance between the antennas should ensure operation in the far field, i.e. at a distance d, or at least $2d_{ff}$, being twice the distance of the border distance for the far field d_{ff}, expressed in the relationship (48), where D_A is the largest linear

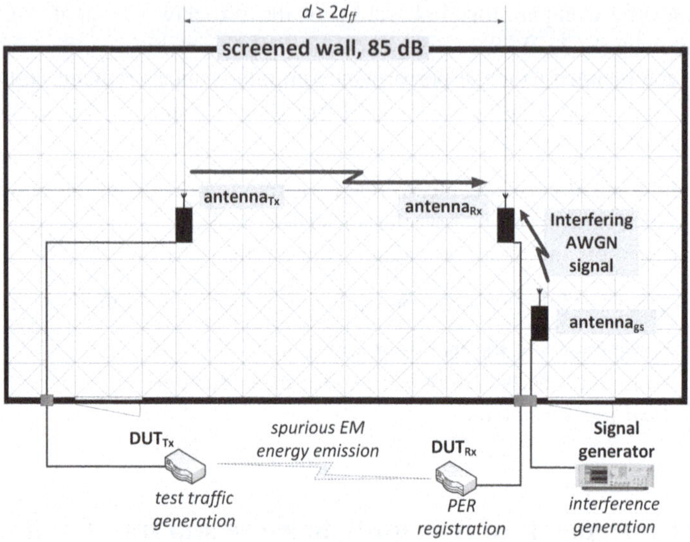

Fig. 6.1 A schematic of the measurement setup for immunity/performance tests of LTN systems against EM interference, with the use of the anechoic chamber

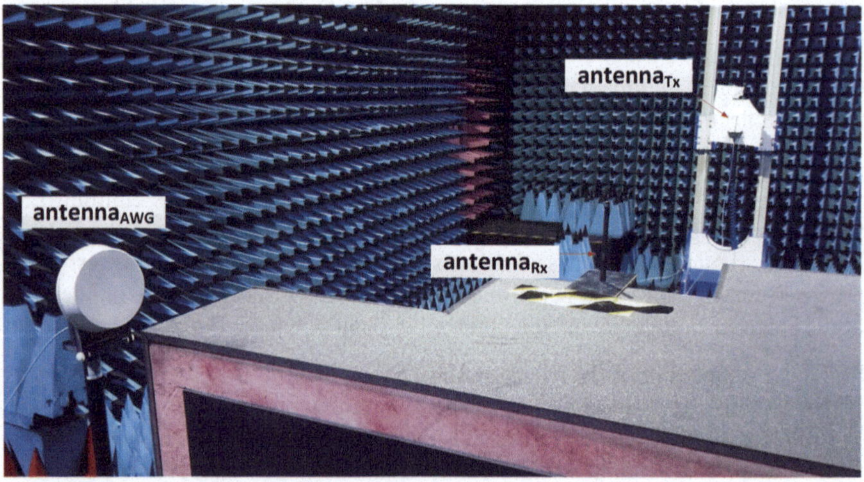

Fig. 6.2 Measuring set for immunity/performance tests of LTN systems for EM interference, in the real WUST ECL anechoic chamber

dimension of the antenna in [m], whereas f is the highest frequency component of the radiated radio signal, expressed in [MHz]. Respective orientation of the antennas should allow maximizing the total energy gain in the energy balance. Due to the fact that LTN systems are by definition narrowband, it is possible and desirable to use omnidirectional antennas (dipoles or monopoles).

$$d_{ff}\left[m\right] = 2\frac{D_A^2}{\lambda} = \frac{D_A^2 \cdot f}{150} \qquad (6.1)$$

2. **Location of the interfering antenna**: the third antenna in the measuring set is the antenna 'antenna$_{gs}$' emitting interference of controlled power level, generated by a signal generator, in the direction of the receiving antenna. Due to the fact that it is the element of the system that is used to control the carrier-to-noise ratio (CNR) as 'seen' by the (DUT$_{Rx}$) receiver, it should be an antenna with directional radiation characteristics, such as a horn or a panel antenna, with the main beam aimed at the receiver antenna ('antenna$_{Rx}$').
3. **Transmitter**: in the preparatory stage, the transmitter module should be electromagnetically isolated from the receiver, e.g. by placing it in a shielded container (cassette, shielded case, etc.). Proper isolation of both systems is intended to ensure that their communication will take place only between the two antennas and not, for example, due to the parasitic leakage of EM energy (marked as 'undesired emission of EM energy' in Fig. 6.1). These outflows may have source, e.g. in antenna connections usually placed either directly on the circuit board or extruded of its housing, to which the cable outputting the signal to the antenna is connected.
4. **Receiver**: at the preparatory stage, the receiver system should be electromagnetically isolated from the transmitter, e.g. by placing it in a shielded container (cassette, shielded case, etc.).
5. **Interference system**: a set consisting of a signal generator possessing the function of generating signals with AWGN spectral characteristics and an adjustable attenuator with a large dynamic range and a 1-dB step. In the measurement results quoted further in the chapter, Tektronix AWG 7000 arbitrary waveform generator was used and Vaunix LDA-602 programmable attenuator with the attenuation range of 0–63 dB (with a minimum step of 0.5 dB).

Stage II: Tuning of the Hardware Settings for Measurements

The aim of this stage is to find a dynamic range for the CNR (CNIR) ratio of the IoT system under investigation, within which it will undergo a transition between two of its extreme operational states. The first of these states marks the ability to work without losing packets regardless of the selected Operating Mode, or 'OM', determined most often by the available so-called modulation-coding schemes or 'MCS'. The second extreme condition is the total inability to work properly, resulting in $PER = 100\%$, even for the OM with minimal requirements with respect to CNR. The concept of dynamic range is depicted in Fig. 6.3, where OM_1 represents the modulation-coding combination with the lowest requirements, whereas OM_N represents a combination of the highest requirements with respect to CNR (CNIR). The transition between PER = 0% and PER = 100% for OM_1 occurs at the lowest values of CNR (CNIR) compared to other OMs.

The diagrams in Fig. 6.3 show typical view of CNIR curves transiting between the states: ON (PER = 100%) and OFF (PER = 0%) of the system. The curves are

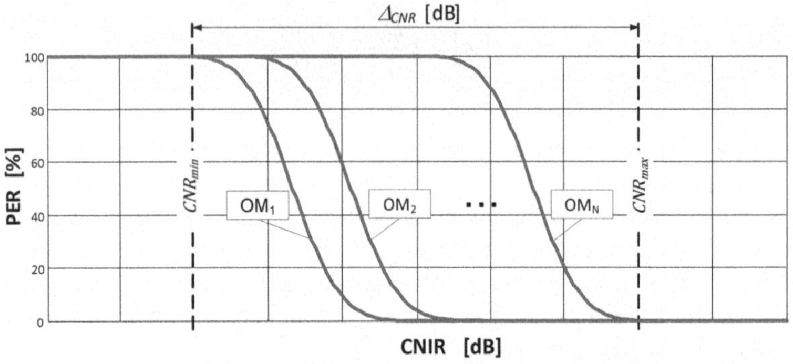

Fig. 6.3 The dynamic range concept of the Δ_{CNR} of the tested IoT system

in fact inverted profiles of the cumulative distribution function for correct reception. The dynamic range CNR (CNIR) marked as Δ_{CNR} is given by Eq. (6.2), where N signifies the TO index with the highest-quality requirements relative to the signal-to-noise ratio. It should be noted here that the concept of the operating mode is not standardized and depends on the operational specifics of a particular system. Most often it is a combination of a modulation and coding efficiency R, i.e. MCS, e.g. {QAM; 1/2}. An 'OM', however, can also be determined by other parameters, such as the channel width (BW), the spreading factor (SF) or the number of repetitions (as in the NB-IoT system). Regardless of which parameters define the operating modes of a given system, they always define a certain set of levels between which the switching (usually automatic) takes place in response to specific noise interference conditions. This behaviour is often referred to as the automatic rate fallback (ARF). When performing performance-immunity tests, however, this function should be deactivated, in order to coerce only two specific OMs: the one with the lowest (OM_1) and the one with the highest (OM_N) CNR requirements.

$$\Delta_{CNR}[dB] = CNR_{\max} - CNR_{\min} \tag{6.2}$$

The procedure described below for determining Δ_{CNR} is also shown schematically in Fig. 6.4:

1. **Transmitter**: in order to ensure the most optimal operating conditions, it is recommended to set the transmitter to work with the highest possible power level P_{DUT}.
2. **Transmitter, receiver**: at the tuning stage, before starting the proper tests, measurements should also be carried out to verify the effectiveness of measures taken to isolate the transmitter and receiver systems, consisting of:
   (a) Disconnecting both antennas ('antenna_{Rx}' and 'antenna_{Tx}') and then terminating the cables supplying the signal with impedance matching loads.
   (b) Coercing an OM with the lowest requirements relative to the CNR, i.e. OM_1. If in all three attempts, 100 packets in each, PER is 100%, it can be assumed

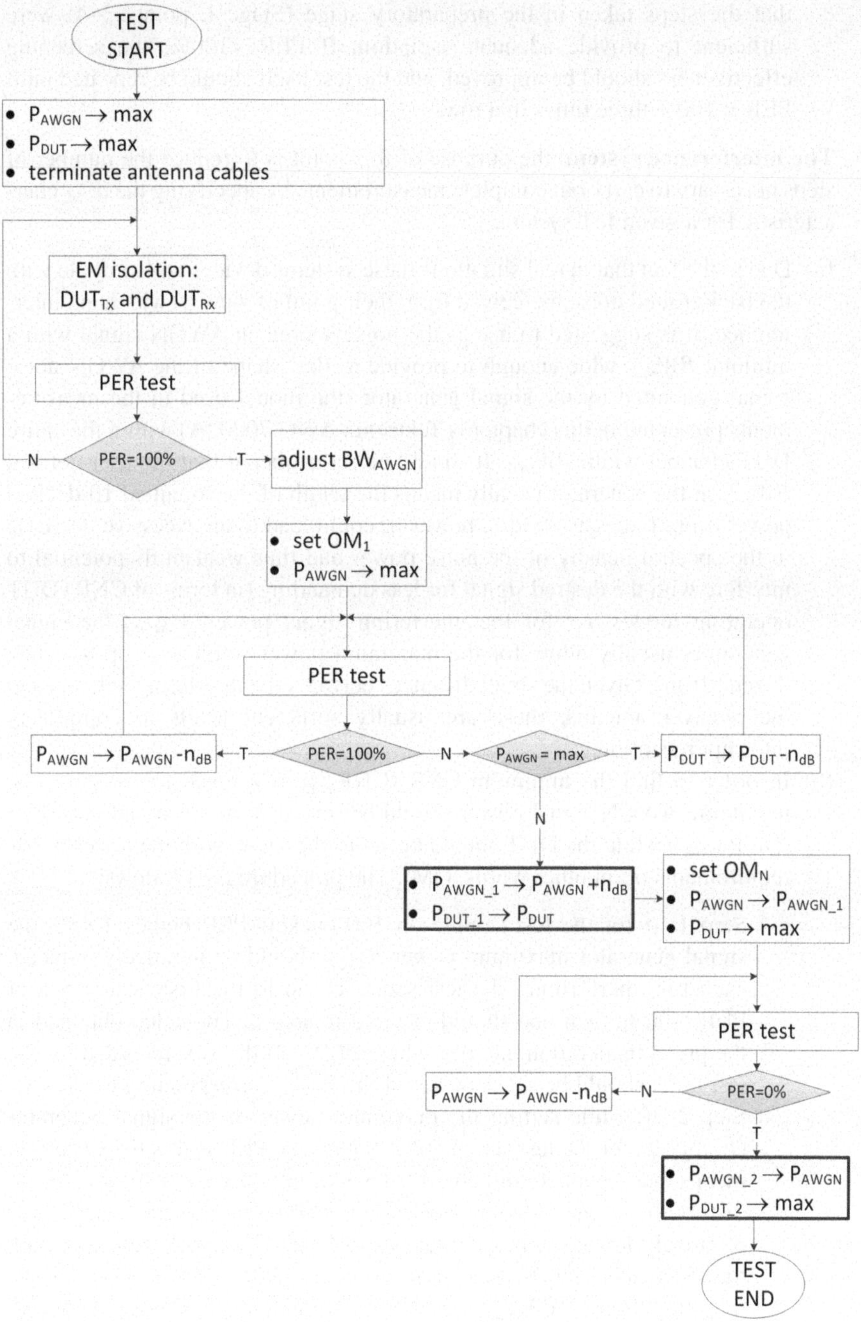

Fig. 6.4 A schematic procedure for determining the dynamic range of (Δ_{CNR}) measurements

that the steps taken in the preparatory stage (Stage I, points 3–4) were sufficient to provide adequate isolation. If PER <100%, the screening effectiveness should be improved, and the test itself should be repeated until PER = 100% three times in a row.

3. **The interference system**: the purpose of this point is to reduce the number of steps necessary to carry out complete measurements by specifying the Δ_{CNR} characteristic for a given IoT system.

   (a) Due to the fact that in real situations these systems devices will struggle with the background noise, perceived from their point of view as wideband interference, it is suggested to use as the noise source an AWGN signal with a minimal BW_{noise} wide enough to provide a 'flat' shape of the AWGN noise signal generated by the signal generator (the model used in the measurements presented in this chapter is Tektronix AWG 70002A) within the entire DUT channel width BW_{DUT}. It should be remembered that the user-defined BW_{noise} in the generator usually means the width of the so-called 10-decibel power drop. Excessive width, however, could lead to an excessive decrease in the spectral density of the noise power and thus weaken its potential to interfere with the desired signal for less demanding (in terms of CNR) DUT operating modes. As for the interfering signal power, P_{AWGN}, the signal generators usually allow for the maximum power settings of up to a few dozen [dBm]. Given the small distances between the interfering antenna and the receiver antenna, these are usually sufficient levels to completely interrupt reception.

   (b) In order to find the minimum CNR (CNR_{min}) of a given IoT system, the maximum AWGN signal power should be forced from the signal generator (i.e. P_{AWGN}), while the DUT should be set to the mode with the lowest CNR requirements or, in other words, OM_1. The procedure is as follows:

      – Step 1: if for the test sample on 100 packets PER equals 100%, the signal generator maximum power P_{AWGN} should be iteratively reduced, each time performing the test sequence up to the first occurrence of PER<100%; then one should proceed to step 2. The value obtained in the previous iteration, i.e. one where PER = 100% was recorded for the last time, should be set as power of the P_{AWGN} corresponding to CRN_{min}.
      – Step 2: if, while setting the maximum power of the signal generator (P_{AWGN_max}) for the test sample of 100 packets, PER is less than 100%, it means that the maximum possible level of interference is still too low to disturb the tested system regardless of its OM. One should then begin to iteratively lower the transmitter power level, P_{DUT}, each time repeating the test sequence up to the first occurrence of $PER = 100\%$, which means that the desired CNR_{min} for the measurements has been achieved. Lowering the output level of the transmitter can be software controlled directly through the device control panel or by means of a regulated attenuator attached to the output of the DUT transmitter.

- The resulting set of boundary parameters for CNR_{min} is therefore defined by the combination $\{P_{AWGN\_1}; P_{DUT\_1}\}$.

(c) In order to find the maximum CNR (CNR_{max}) of a given IoT system, the DUT should be set to the OM_N mode (i.e. one with the highest requirements for CNR/CNIR). The power of the output signal from the signal generator should be P_{AWGN_1}, whereas the DUT_{Tx} should work at the maximum power (P_{DUT_max}), remembering that if in point 3b the DUT_{Tx} output had to be additionally attenuated ($P_{DUT\_1} < P_{DUT\_max}$), this attenuation should now be turned off (setting the value to 0 dB). The procedure is as follows:

- Examine PER with the 100-packet series, iteratively reducing the P_{AWGN} level each time after each series, until the PER result is 0%. Initial changes in the AWGN signal level can take place in any step width, but it is proposed to take a 5 dB step for a quick preliminary estimation of the interference level, for which PER = 100%. Having achieved this point, i.e. the exact value of interference level at which PER = 0%, may be arrived at for the first time by finer changes in the AWGN signal level, e.g. every 1 dB. The expected, typical dynamics of the systems operation is about 20–25 dB.
- The resulting set of border parameters for CNR_{max} is thus defined by the combination of $\{P_{AWGN\_2}; P_{DUT\_2}\}$.

(d) The dynamic range for Δ_{CNR} measurements is, therefore, the difference [dB]: CNR_{max}-CNR_{min}.

Stage III: PER Measurements

The procedure described below that leads leading to the finding of Δ_{CNR} is also shown schematically in Fig. 6.5:

1. **Interference system**: initial settings of the interfering setup should be appropriate for the state of CNR_{min} or in other words $\{P_{AWGN\_1}; P_{DUT\_1}\}$ and then:

   - Step 1: if $P_{DUT\_1} = P_{DUT\_max}$ go to step 2. If $P_{DUT\_1} < P_{DUT\_max}$ increase P_{DUT} iteratively first, until P_{DUT_max} level has been reached, then go to step 2. The size of the suggested jump is 1 dB, which allows to obtain 'smooth' transition profiles between $PER = 100\%$ and $PER = 0\%$. As will be demonstrated in the next chapter devoted to the measurement results of exemplary IoT systems, this transition usually takes place on the CNR span, approximately 6–8 dB wide.
   - Step 2: reduce the P_{AWGN} level iteratively until the $PER = 0\%$ level in the receiver is reached.

2. **Transmitter, receiver**: both devices should be switched between available operational modes, in the increasing order with relation to the requirements for CNR (CNIR), i.e. OM_1, OM_2, …, OM_N. After each switch, a 100 transmission test should be performed for each of the settings of the interfering system and the

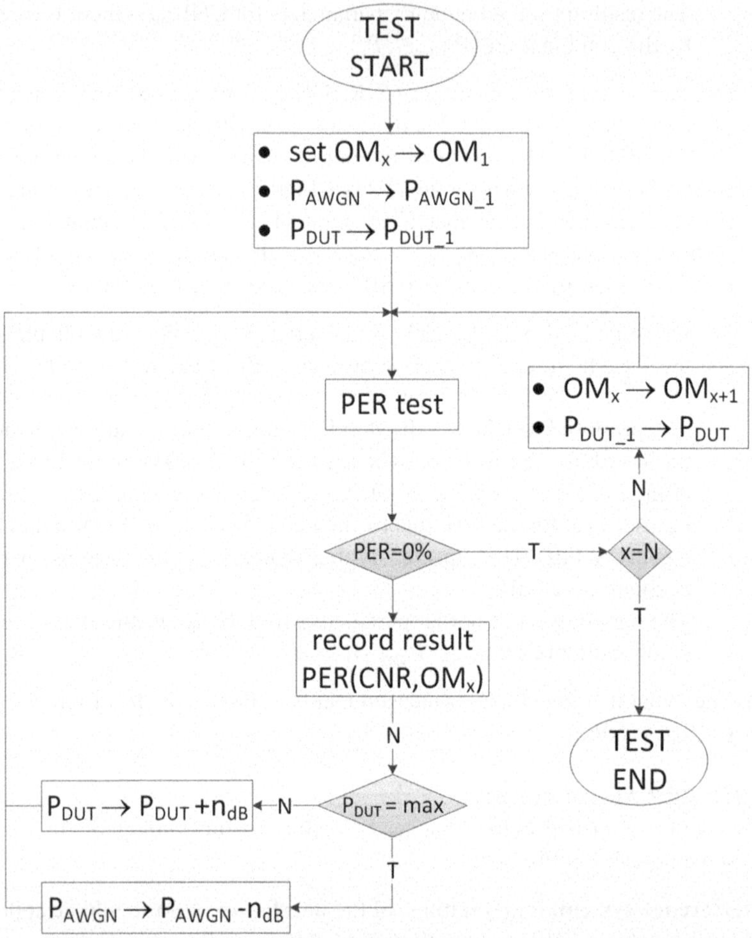

Fig. 6.5 Diagram of the PER measurement procedure using the anechoic chamber

resultant *PER* be registered. The dynamic range of measurements, defined in Stage II, should allow for the transition between extreme PER states, whereas the further proposed settings of the interfering system serve the purpose of ensuring that this transition occurs between *PER* = 100% and *PER* = 0%. Achieving this status completes the test for x-th operational mode (OM_x).

3. The result of such a procedure should be a set of *N* transition profiles showing *PER* as a function of CNR, as shown in Fig. 6.3.

6.3 Measurements of Immunity to Extremely Multipath Propagation

The EM signal emitted from the antenna propagates through multiple interactions (i.e. reflections, obstacles, diffraction at the edges or scattering) with elements of the propagation environment. As a result, the signal reaching the receiver will take the form of a time series of signals with different amplitudes of the electric field intensity (*E*), phases (θ) and delays (τ), i.e. of multipath components, as illustrated in Fig. 6.6. At the receiving point, they are vector added, and, due to the random nature of their phase differences, the resultant signal has a time-diminishing character, and the fluctuations in the envelope of such a signal are called small-scale decay. The amplitude probability distribution of these decayed echoes is usually modelled using the Gaussian, Nakagami-Rice and Rayleigh distributions [113], [114], depending on the proportional contribution of the direct line-of-sight (LOS) and the indirect non-line-of-sight (NLOS) components. The mathematical model of such a channel can thus be presented in the form of a linear, time-varying filter with an impulse response $h(t,\tau)$, according to formula (6.3). Time variability is sometimes caused by temporary changes in the geometry of the propagation environment, e.g. due to movement of people or objects moving or due to shifting the position of the transmitter or receiver. Another form of multipath channel representation is the Power Delay Profile (PDP), given by Eq. (6.4).

$$h(t,\tau) = \sum_{i=0}^{N(\tau)-1} E_i \delta(t-\tau_i,\tau) e^{j\theta_i(t,\tau)} \tag{6.3}$$

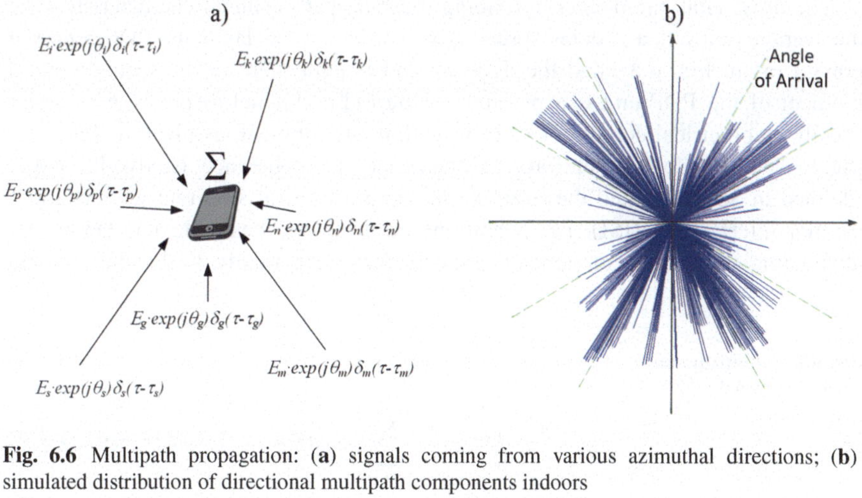

Fig. 6.6 Multipath propagation: (**a**) signals coming from various azimuthal directions; (**b**) simulated distribution of directional multipath components indoors

$$PDP(t,\tau) = \sum_{i=0}^{N(\tau)-1} P_i \delta(t-\tau_i,\tau) \tag{6.4}$$

where

- τ_i – delay of the *i*-th multipath component [s]
- τ – observation time [s]
- E_i – electric field intensity of the *i*-th multipath component [V/m]
- θ_i – intensity phase of the *i*-th multipath component of the electric field [rad]
- P_i – the power of the *i*-th multipath component [W]
- $\delta_i(\cdot)$ – the Dirac delta impulse representing the *i*-th multipath component
- $N(\tau)$ – number of multipath components in the course of the observation τ

The frequency response of the *H(f)* channel, also called colloquially – though not entirely correctly – the channel transmittance, thus takes on the form of the Fourier transform $h(t,\tau)$, in accordance with formula (6.5). Both the impulse response and the frequency response thus exhaustively describe the status of the multipath radio channel. It is sufficient to know one, since the other can be obtained using Fourier transform or its inverse. Assuming the AWGN radio channel, its output signal *y(t)*, also being the input signal of the receiver, is given by the formula (6.6), where x *(t)* in *(t)* are the input signal (transmitted from the transmitter) and AWGN, respectively, as in Fig. 6.7.

$$H(f) = \int_{-\infty}^{\infty} h(t,\tau) e^{-j2\pi ft} dt \tag{6.5}$$

$$y(t) = \int_{-\infty}^{\infty} x(t) h(t,\tau) d\tau + n(t) \tag{6.6}$$

The most vital parameters describing the multipath channel characteristics are the average delay τ_m and delay spread τ_{RMS}. The average delay is the PDP centre of gravity (as in Fig. 6.8), and the delay spread is equivalent to the second central moment of the PDP and is a measure of the radio channel dispersiveness under specific geometrical conditions of propagation environment, as given in Table 4.1 (Sect. 4.3). It allows, therefore, to determine the coherence bandwidth BW_{coh} (defined in Sect. 4.3) and the resulting maximum transmission rate without intersymbol interference (ISI), i.e. conditions ensuring the desirable flat fading, by attenuating all spectral components of the useful signal equally. Both values, τ_m and

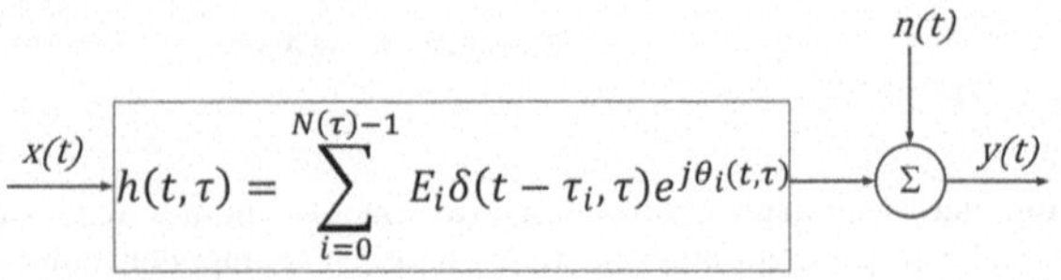

Fig. 6.7 Multipath radio channel with AWGN noise model

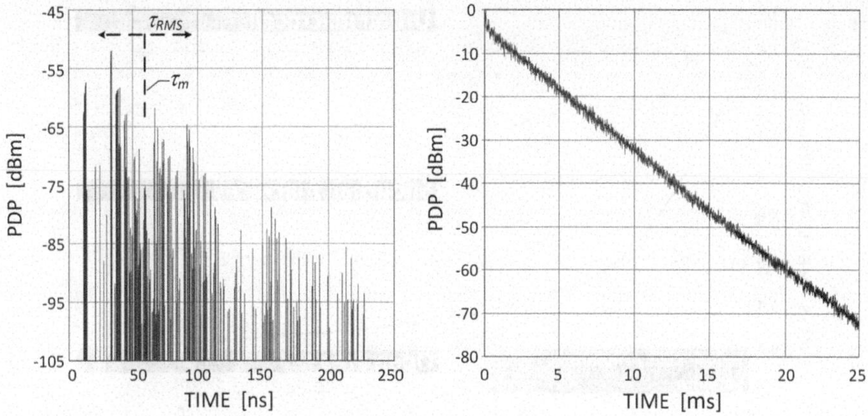

Fig. 6.8 Examples of PDP Power Delay Profiles: simulated for the interior of the building (left); measured for an unloaded reverberation chamber located at WUST ECL (right, based on [115])

τ_{RMS}, knowing the PDP profile components, can be calculated with formulas (6.7) and (6.8), respectively.

$$\tau_m = \frac{\sum_i \tau_i \cdot P(\tau_i)}{\sum_i P(\tau_i)} \tag{6.7}$$

$$\tau_{RMS} = \sqrt{\frac{\sum_i (\tau_i - \tau_m)^2 \cdot P(\tau_i)}{\sum_i P(\tau_i)}} \tag{6.8}$$

In order to measure the influence of the multipath propagation on the operation of IoT systems, it is recommended to use a reverberation chamber with a Power Delay Profile (PDP) known via measurements. The tests described in the next chapter are based on the results obtained from previous delay spread τ_{RMS} measurements performed for various stages of the chamber load. Each stage was characterized by a different number of absorbers reducing the chamber reflective potential. As a result, the τ_{RMS} profile was arrived at, shown in Fig. 6.9 and previously described in [115], which represents the chamber response to the gradual addition of absorption panels, starting with the unloaded chamber condition (with $N = 0$ panels) and ending with the overloaded state ($N = 12$ panels), for which any further loading did not cause significant changes in τ_{RMS}. The extremely multipath conditions correspond to $\tau_{RMS} \approx 1.55$ µs, which is the value situated in the range typical for urban macrocells (1–3 µs acc. to [93]).

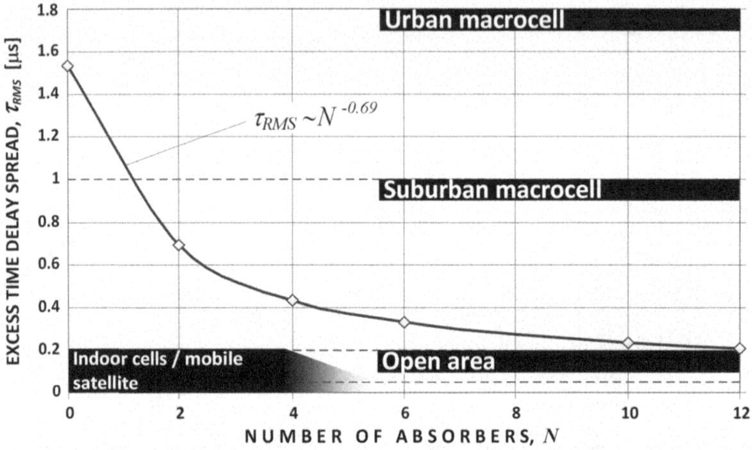

Fig. 6.9 Power Delay Profile (PDP) measured in the reverberation chamber located WUST ECL (based on [115])

Stage I: Preparation of the Measurement Setup

1. **Arrangement of the transceiver system antennas**:

   (a) In order to eliminate the influence of the tested DUT devices, e.g. by their undesirable energy radiation from onboard electronics, only antennas of these devices should be placed in the chamber, marked in Figs. 6.10 and 6.11 as '$antenna_{Tx}$' and '$antenna_{Rx}$', attached to their respective DUTs by means

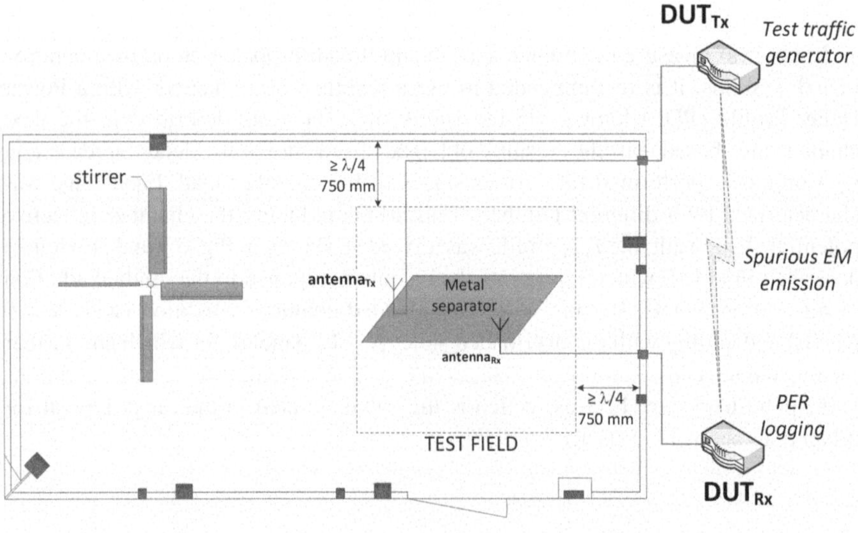

Fig. 6.10 A schematic of measurement testbed measuring the impact of extreme multipath propagation on the operation of the IoT system WUST ECL reverberation chamber

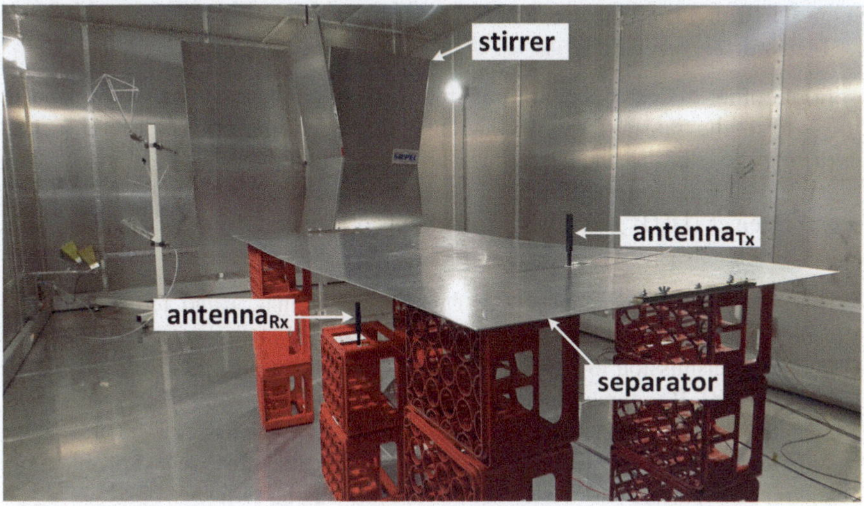

Fig. 6.11 The measurement setup for testing the immunity/performance of LPWAN systems to the multipath phenomenon in the reverberation chamber at WUST ECL

of low-loss cables with the minimum length required to operate the equipment outside the chamber during measurements. These cables should be led outside the chamber through well-shielded feedthroughs in the chamber wall.

(b) In order to eliminate the direct component, both antennas, during PER measurements, should be separated by a conductive separator (e.g. a metal plane), which should create the signal reception conditions characteristic of the Rayleigh channel (desirable and typical for resonance cavities acc. to [116]), i.e. one in which the arriving components are solely due to reflections and diffractions.

(c) Both antennas should be placed in the test field of the reverberation chamber, defined as the area distant from the walls of the chamber by at least $\lambda_{maks}/4$. It is equivalent to 75 cm in the case of WUST ECL, whose lowest usable frequency f_L determined for this chamber equals 100 MHz (and therefore $\lambda_{maks} = 3$ m).

(d) During the preparatory stage, the transmitter (DUT_{Tx}) and receiver (DUT_{Rx}) systems should be electromagnetically isolated from each other, e.g. by placing them in shielded casings.

2. **Stirrer**:

(a) The reverberation chamber should be equipped with a stirrer that facilitates a homogeneous distribution of the electric field in its interior.

(b) The chamber control system should enable the stirrer to work in a continuous rotation mode, with a user-defined angular rotation speed [degrees/second].

Stage II: Tuning the Hardware Settings for Measurements

1. **Transmitter, receiver**:

   (a) In order to coerce the best possible working conditions, it is recommended to set the transmitter to work with the highest possible power level P_{DUT}.

   (b) Before starting the tests, measurements should be made to check the effectiveness of the measures taken to isolate the transmitter (DUT_{Tx}) and receiver (DUT_{Rx}) systems situated outside the reverberation chamber. The check consists in:

- Disconnecting both antennas ('antenna$_{Rx}$' i 'antenna$_{Tx}$') and terminating the cables supplying the signal with impedance-matched loads.
- Forcing transmission to OM_1 with the lowest requirements relative to the CNR level (CNIR). If in three trials, 100 packets in each, PER will be 100%, it can be assumed that the measures taken in the preparatory stage (Stage I, point d) were sufficient. If PER<100%, it is necessary to improve the effectiveness of shielding and repeat test tests until PER = 100% three times in a series.

2. **Stirrer**:

   (a) In accordance with IEC 61000-4-21 Part: 4 [117], in order to obtain a properly reliable response of the reverberation chamber to the signal stimulation, tests should be carried out for 12 different angular stirrer positions, in increments of 30°, which is equivalent to a single full rotation. The number of positions defined in this way guarantees that the measurement uncertainty of the average electric field intensity will be 5 dB at the 95% confidence interval and is valid for the frequency range $6{\cdot}f_L$–$10{\cdot}f_L$. In the case of WUST ECL, it corresponds to the range 600 MHz–1000 MHz, in which the centre frequencies f_c of operational channels for most LTN systems are located. It is worth noting that the choice of 12 stirrer positions is a compromise between the measurement accuracy, the measured uncertainty of measurements and the duration of tests: in the case of measurements taken at a single position, uncertainty in the mean field value can be as great as 30 dB compared to the actual one, though in favour of 12 times shorter measurement time. For 100 revolutions, in turn, this uncertainty drops to the level of 2.4 dB but at the expense of an eightfold extended measurement time relative to the case with 12 revolutions.

   (b) The angular speed of the stirrer should be set to the maximum value.

Stage III: PER Measurements

The procedure for performing PER tests in the reverberation chamber described below is also shown schematically in Fig. 6.12.

1. **Transmitter, receiver**: both devices should be switched between the available operational modes, in increasing order in relation to the requirements for CNR (CNIR), i.e. in the sequence OM_1, OM_2, …, OM_N. After each switch, a 100-packet transmission test should be performed for both of the stirrer modes presented below, registering the resulting PER:

(a) The immovable stirrer mode, i.e. a 'static mode' consisting in performing a separate PER test for each of at least 12 angular positions of the stirrer, marked 'Pos_i' in the figure (where 'i' is the index of the given item).
(b) Stirrer mode, i.e. a 'dynamic mode', consisting in performing the PER test during the stirrer continuous rotation (item marked 'Pos_c' in the figure).

2. **Stirrer**: the maximum stirrer speed can result in a full turnover for the operating modes offering the lowest throughput prior to completion of the full 100-packet test sequence. In order to avoid this, either of two activities should be done:

(a) Set the stirrer into a continuous operation (allowing for many turns in a row).
(b) Ensure that each time one full revolution is completed, the next one is initiated immediately, until the PER test sequence has ended.

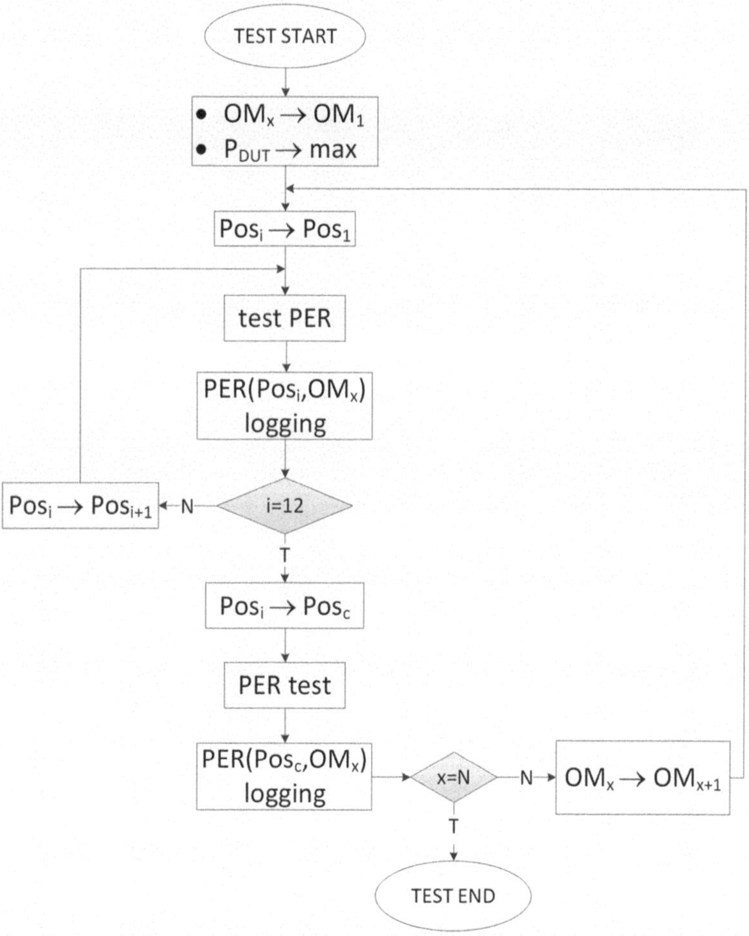

Fig. 6.12 A diagram illustrating PER measurement procedure using the reverberation chamber

Chapter 7
LPWAN/UNB IoT Systems Performance Investigations

The results of performance measurements presented in this chapter were obtained for a representative group of two LPWAN IoT systems: LoRa, Weightless (-P) and SigFox belonging to the UNB group. The measurements were made using the methodology presented in Chap. 6, broken down into parts related to:

- The immunity to interference
- Extremely multipath propagation

7.1 LoRa System

7.1.1 Immunity to Noise and Interference

In the case of the LoRa system, two types of interfering sources can be distinguished: other LoRa transmitters and interference from other system transmitters. The threat of the first type occurs in the situation of simultaneous transmission to/from two or more devices using the same set of parameters $\{f_c, BW, SF\}$, because this configuration does not ensure orthogonality of signals, as discussed in Sect. 5.1.5.3. As a result, such signals become inseparable in the detector.

Due to the fact that in real situations the LoRa system devices will struggle with background noise, viewed as the wideband interference, the noise source described in this chapter was the 1000 kHz AWGN signal generated by the arbitrary signal generator Tektronix AWG 70002A with an initial output power of 7 dBm, selected so that in the process of switching between individual *SF* values, a transition from $PER = 0\%$ to $PER = 100\%$ could be observed. This power was then reduced with a 3-decibel step-down to −8 dBm, observing the impact of the successively increasing SNR ratio on the resultant Packet Error Rate, in order to obtain a quantitative validation of the recommendations contained in Sect. 5.1.5.3, regarding the

K. Staniec, *Radio Interfaces in the Internet of Things Systems*,
https://doi.org/10.1007/978-3-030-44846-2_7

proposed optimal settings of *BW*, *SF* and *R* parameters, depending on the quality of propagation and interference conditions.

The measurement setup was arranged in the anechoic chamber of the Electromagnetic Compatibility Laboratory (ECL) of the Wrocław University of Science and Technology in the configuration shown in Fig. 6.1. The LoRa DUT devices used in the research were "Eiger platform" modules based on the Semtech SX1276 chipset. Analysis of the data collected during the measurements has led to three basic conclusions regarding the response of the LoRa system to interference:

Conclusion 1: as expected, the operational mode with $BW = 500$ kHz channels is the least resistant; the lowest *PER* was available for high values of the spreading factor, although its value did not exceed 10% only for $SF = 12$. As shown in Sect. 5.1.5.3, such a wide channel width provides the highest transmission rate, as shown in formula (36), but at the same time increases the noise power, thus lowering the SNR ratio. In the studied cases, its value was at least −19 dB (for $BW = 500$ kHz, $SF = 12$ i $P_{gen} = 7$ dBm), successively increasing by 3 dB with each change of the channel width to twice narrower, or, respectively, −16 dB and −13 dB for, respectively, $BW = 250$ kHz and 125 kHz (and $SF = 12$). The results quoted here allow one to observe a certain practical regularity in PER behaviour as a function of SNR. Namely, *SF* modes with PER close shift towards lower values (e.g. $SF = 12$ towards $SF = 11$ for $BW = 500$ kHz or $SF = 10$ towards $SF = 9$ for $BW = 125$ kHz) with a step change of one *SF* per every 3 dB of SNR reduction (Fig. 7.1).

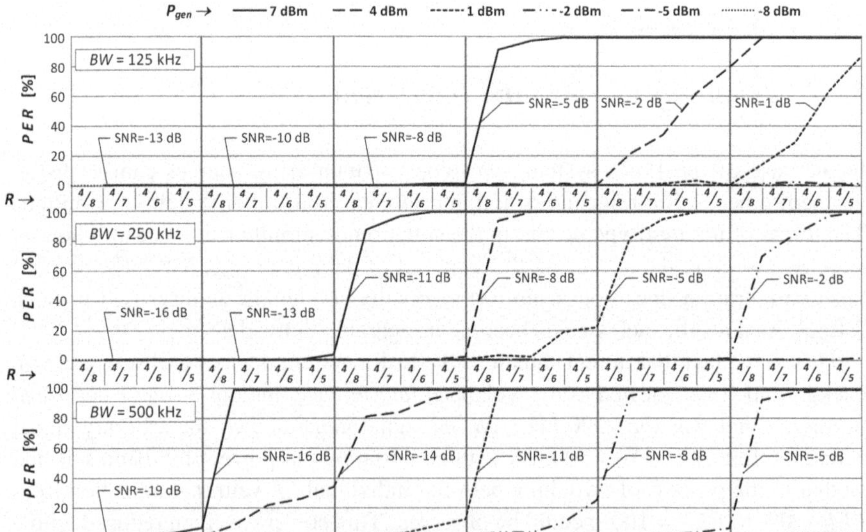

Fig. 7.1 Packet Error Rates measurement results for LoRa under controlled interference conditions [117]

Conclusion 2: a similar effect to the previously described can be observed with respect to *BW* changes, namely, each time a double reduction in *BW* (i.e. from 500 kHz to 250 kHz and then to 125 kHz) also reduces the *SF* index for which PER achieves low values:

- For *BW* = 500 kHz, a low PER is available for *SF* = {12}.
- For *BW* = 250 kHz, a low PER is available for *SF* = {11, 12}.
- For *BW* = 125 kHz, a low PER is available for *SF* = {10, 11, 12}.

Conclusion 3: the effect of changes in the protection coding efficiency can be observed by altering its value from *R* = 0.5 to 0.8 within each *SF*. The coding, as expected, causes *PER* to increase with increasing *R*, within a given *SF*, from the lowest level of *R* (providing the greatest protection) on the left edge of the *SF* area (where *R* = 0.5) to the highest level on the right edge (where *R* = 0.8). The noticeable effect of this observation is a fluent profile of transition between extreme PER values (i.e. between 0% and 100%) accompanying the changes in *SF*.

7.1.2 Immunity to the Multipath Propagation

As with the interference immunity measurements described in Sect. 7.1.1, LoRa devices were tested for under multipath conditions at three possible values of *BW* = {125 kHz; 250 kHz; 500 kHz}, six values of the spreading factors *SF* = {7–12} and two extreme values of the coding efficiency *R* = {0.5; 0.8}.

The analysis of the results presented in Fig. 7.2 allows to formulate the following conclusions:

Conclusion 1: similar to the LoRa interference immunity tests, higher degrees of the signal spread factor *SF* values result in greater immunity to the strongly multipath propagation in the Rayleigh channel. Evidently, for the three largest *SF* values (12–10), the system is practically completely insensitive to the multipath propagation, yielding *PER* = 0%, regardless of the *BW* used. Beginning from *SF* = 9, for certain stirrer positions, singular high *PER* values appear reaching 60% for *BW* = 125 kHz, ending with *SF* = 7 where *PER* in excess of than 80% was recorded for that *BW*.

Conclusion 2: the assessment of *BW* impact on susceptibility to the multipath propagation does not seem so obvious, since assuming that a mean *PER* is still a reliable indicator observed for all stirrer rotations, smaller *BW* seems to be more sensitive, yielding *PER* of 6.4% and 31.3%, respectively, for *SF* = 10 and 11. Whereas for channels with *BW* = 500 kHz and with the same values of spreading factors, *PER* levels are rather negligible (from 0.5% to 2%). In turn, the results obtained for *SF* = 7 seem to agree again with the assumption of the least immunity of channels with *BW* = 500 kHz, because PER obtained there significantly exceed the values measured for other channel widths, namely, between 47% and 56%, respectively, for *R* = 0.5 and 0.8.

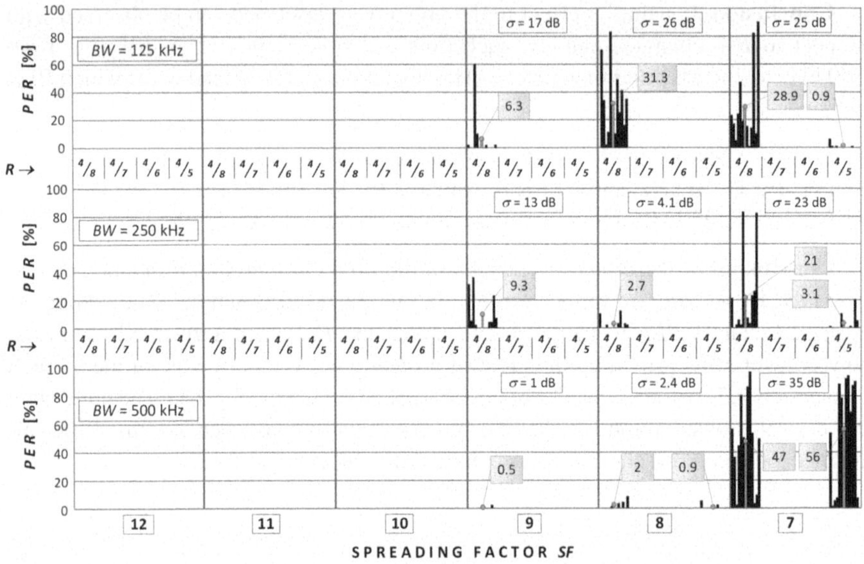

Fig. 7.2 Measurement results for the LoRa system in the reverberation chamber [117]

Conclusion 3: the measured responses of the LoRa system to multipath propagation reveal a considerable unevenness in relation to the different positions of the stirrer. In order to quantify this phenomenon, the standard deviation σ was used as a quantifying measure, calculated separately for each *SF*. Clearly, due to significant values of σ reaching 35 dB, the average *PER* values described in the Conclusion 2 are too uncertain to constitute a reliable impact indicator. It can therefore be concluded that the most reliable assessment of the multipath propagation impact is Conclusion 1, which states that regardless of the channel width or coding efficiency, uninterrupted operational modes in LoRa are those with the spreading factor *SF* chosen from within the range 12–10.

7.1.3 Summary

The results of the LoRa system immunity measurements presented in this chapter confirm the general planning principle that under conditions of increased electromagnetic interference, lower *BW* channel widths and higher *SF* scattering factors should be used to reduce compliance. A linear dependence between *BW* and *SF* is demonstrated in Sect. 5.1.5.2, showing that per every twofold reduction in *BW*, the spreading factor *SF* index can be decreased by '1' (e.g. from *SF12* to *SF11* during the transition from *BW* = 500 kHz to 250 kHz), while *SF* = 12 proved insensitive to interference even at the largest channel width.

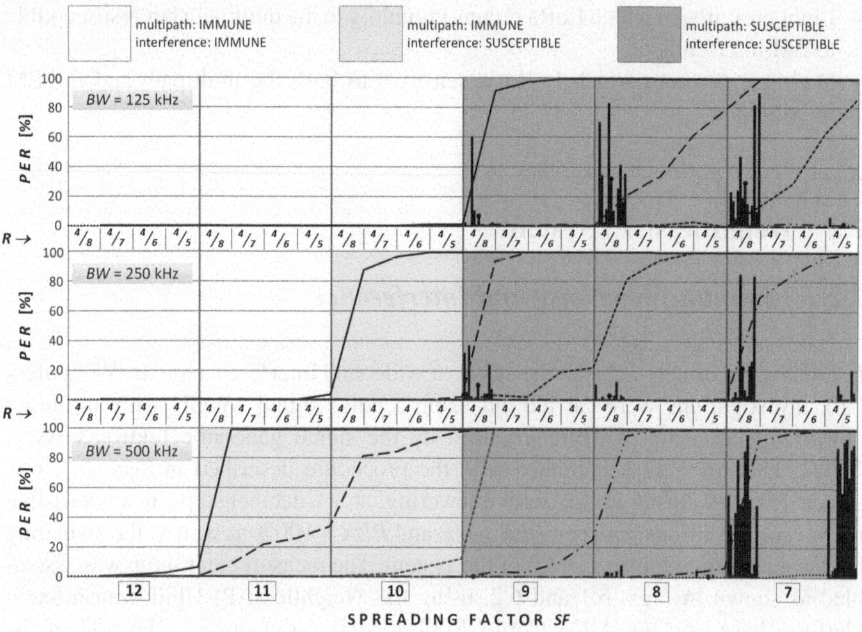

Fig. 7.3 A compilation of areas of immunity and susceptibility to interference and multipath propagation in LoRa system [117]

Table 7.1 A compilation of areas of immunity and vulnerability to interference and multipath propagation in the LoRa system

	Spreading factor, SF					
BW	12	11	10	9	8	7
125 kHz						
250 kHz						
500 kHz						

As for immunity to the multipath interference, in turn, in Sect. 7.1.2, it was shown that regardless of the width of the *BW* channel, LoRa does not show any sensitivity to this phenomenon for modes with *SF* = 12–10, while for the others, it shows a visible susceptibility, though very diverse and difficult to quantify.

Figure 7.3 summarizes the results from both measurement campaigns in a manner leading to further conclusions of practical importance in the LoRa network planning process, to provide immunity to electromagnetic and multipath interference. The obtained results allow to specify three areas of immunity/susceptibility to these two factors, marked in white, light grey and dark grey, summarized for clarity also in Table 7.1:

- White area in which LoRa is completely immune to both EM and multipath interference

- Light grey area in which LoRa retains immunity to the multipath but is susceptible to EM interference
- Dark grey area in which LoRa is sensitive to both the multipath and the EM interference

7.2 Weightless(-P) System

7.2.1 *Immunity to Noise and Interference*

In order to accurately analyse the effect of wideband interference on the Weightless (-P), investigations were carried out using a noise signal AWGN with an output power P_{gen} equal to 14 dBm produced by the signal generator Tektronix AWG 70002. This level, in accordance with the procedure described in Sect. 6.2, was selected so that during its successive lowering, at a 1-decibel steps, it was possible to observe transitions between $PER = 0\%$ and $PER = 100\%$ as well as the switching between operating modes (OMs) of the system. The measurement setup was assembled as shown in Figs. 6.1 and 6.2, using the Weightless(-P) Ubiik transmission platform based on the ARM Cortex-M3@40MHz arrangement. The role of the receiver (DUT_{Rx}) interfered by the AWGN signal was played by the base station (BS), while the client station was used as a test traffic generator (DUT_{Tx}). In order to exclude the possibility of undesired leakage of EM energy between both devices located outside the anechoic chamber, they were properly isolated from each other in shielded cassettes, verifying the shielding effectiveness in accordance with the procedure presented in Fig. 6.4. A set of operational modes consisted of a total of eight combinations of parameters specific to Weightless(-P), namely, {*BW; SF; R*; modulation}, that can take on values quoted in Table 5.11. For the clarity of presentation of results, the *SF* value in the combinations was replaced with the resulting transmission rate R_b. The width of the channel was expressed as 'NB' ($BW = 12.5$ kHz) and 'WB' ($BW = 100$ kHz), thus providing an opportunity to explore the entire range of operational modes, beginning with the most immune combination (with the lowest R_b) of {12.5 kHz; 0.625 kb/s; 0.5; OQPSK} and ending with an *OM* offering the greatest R_b, but the most demanding in terms of radio link quality is defined by the following settings: {100 kHz; 100 kb/s; 1; GMSK}. Due to the considerable time difference needed to send the entire number of test packets, the moment when PER stabilized at a certain value was considered as the end of measurement for a given OM. The analysis of results presented in Fig. 7.4 allows to draw four basic conclusions regarding the influence of the discussed four transmission parameters on the resulting *PER*:

Conclusion No. 1: results from the analysis of differences between NB and WB modes, compiled separately for each modulation. Clearly, regardless of its modulation setting (OQPSK/GMSK), for the narrowband mode (grey waveforms, 'NB'), in order to achieve efficiency comparable to the wideband (black waveforms, 'WB'), the required CNR is lower by approx. 2 dB.

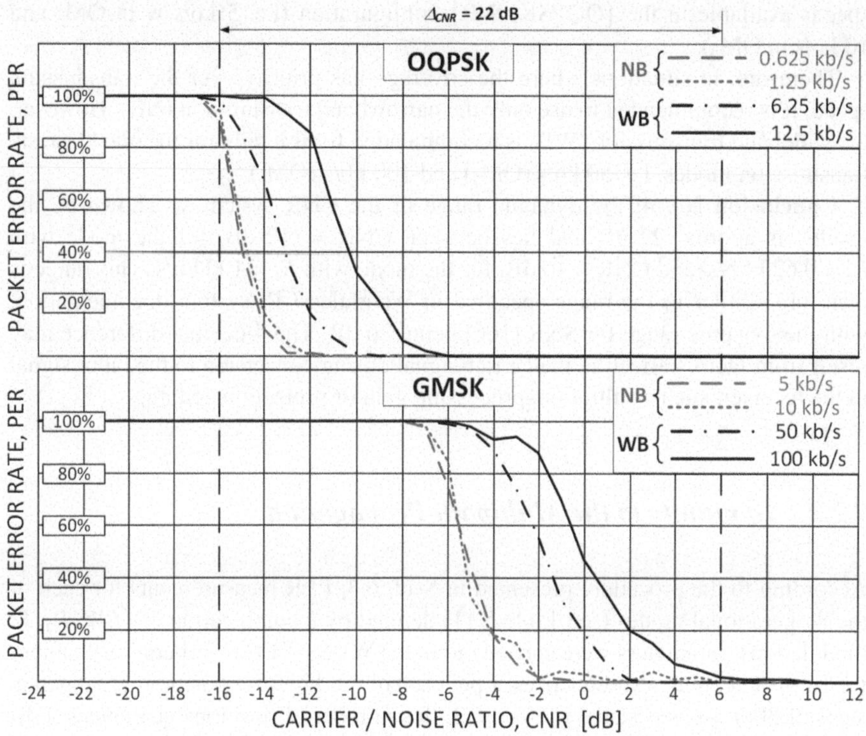

Fig. 7.4 PER measurement results for the Weightless(-P) system obtained in the anechoic chamber

Conclusion No. 2: the transition from OQPSK to GMSK transmission results in an increase in CNR required to sustain a low PER level by approx. $\Delta CNR = 13.2$ dB and 9.3 dB, respectively, for the narrowband and the wideband mode. Due to high demands placed on IoT systems in terms of coverage, this difference will translate into a reduction in the range of Δd given in formula (7.1). Switching from OQPSK to GMSK will, therefore, result in a 4.6- and 3-fold coverage change, respectively, for NB and WB modes, assuming the free space propagation, with the attenuation exponent $n = 2$. Of course, in urban conditions (Chap. 3) where n oscillates around four or indoor values with possibly even greater exponent, this reduction will not be that significant.

$$\Delta d = \frac{d_{\mathrm{OQPSK}}}{d_{\mathrm{GMSK}}} = 10^{\frac{\Delta \mathrm{CNR}}{10 \Delta n}} \left[\mathrm{times}\right] \tag{7.1}$$

Conclusion No. 3: the observed change in the required CNR amounts to 13.2 dB and 9.3 dB, respectively, when changing from OQPSK to GMSK. This time it should be noted that the two highest R_b rates available in the {OQPSK; NB} configuration (i.e. 6.25 kb/s in OM_5 and 12.5 kb/s in OM_6) exceed the two lowest speeds available in the {OQPSK; WB} configuration (i.e. 5 kb/s w in OM_3 and 10 kb/s in OM_4).

Therefore, in situations where the coverage has priority over the transmission speed, it is recommended to use only the narrowband transmission (NB). However, the wideband transmission (WB) is recommended for use only for the two “fastest” transmission modes, i.e. 50 kb/s (OM_7) and 100 kb/s (OM_8).

Conclusion No. 4: the dynamic range of the CNR system, as shown by the results, is approx. 22 dB and lies between CNR = -12 dB, for the mode with $R_b = 0.625$ kb/s, and CNR = 10 dB, for the mode with $R_b = 100$ kb/s. This range is therefore similar to the range specified in Weightless(-P) system documentation, with the dynamic range for SNR [106] being 26 dB. The 4-decibel difference may result from factors affecting the carrier signal during conversion to the input signal in the receiver, such as filtration, processing gain or protection coding.

7.2.2 *Immunity to the Multipath Propagation*

According to the procedure presented in Sect. 6.3, PER measurements for each of the 8 operational modes (see Table 5.11) defined by a combination of {BW; R_b; R; modulation} parameters were carried out in the WUST ELC reverberation chamber for 100-packet long test sequences, 1 per each of the 12 stirrer positions (changed in steps of 30°). As was stated in Sect. 6.3, this number of positions guarantees 5 dB of uncertainty in the average electric field intensity measurements with 95% confidence interval and is valid for the $6{\cdot}f_L$–$10{\cdot}f_L$ frequency range, i.e. in this case 600–1000 MHz, which also includes the centre frequency f_c of the working channel used during the communication of the tested Weightless(-P) system, equal to 863.1 MHz. This series of measurements was called the ‘static mode’ to express the static stirrer position during PER measurement for a given test sequence.

Analogous measurements, in accordance with the methodology presented in figure, were also made for the stirrer rotating in the course of the PER test, hence the name ‘dynamic mode’ to reflect the continuously changing multipath channel due to continuous changes in boundary conditions for the EM wave reflections and diffraction in the chamber. For each operational mode, the measurements in the ‘dynamic mode’ were made three times, assuming their average value for the presentation.

On the summary chart, the results presented in Fig. 7.5 for the ‘static’ (Fig. 7.5a) and ‘dynamic’ (Fig. 7.5b) modes were separated for clarity. In addition, the results obtained for each OM are shown in the form of full bars, 12 and 3 each, which is meant to correspond to the number of individual stirrer positions or the number of PER tests per 100-pack sequence performed, respectively, for the ‘static’ and ‘dynamic’ mode. To the right of the results for each OM, there is also a bar without a filling to indicate the arithmetic mean of full bars describing the results of a given OM.

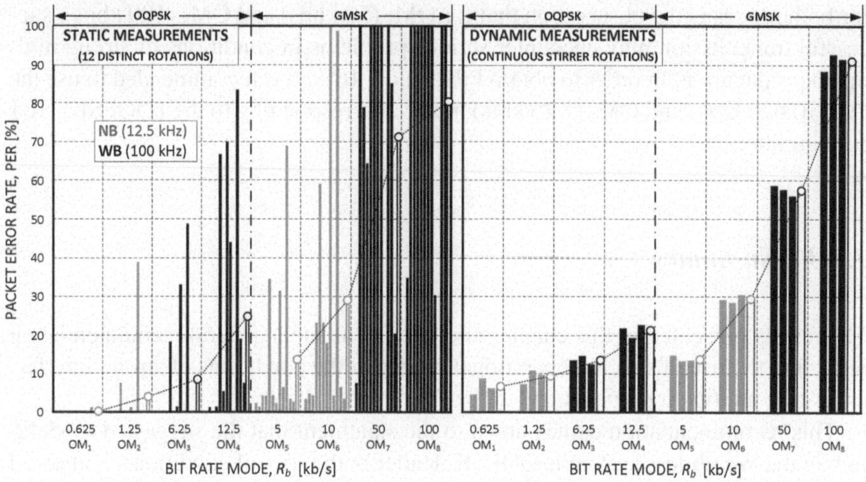

Fig. 7.5 Results of PER measurements for the weightless(-P) system obtained in the WUST ECL reverberation chamber

Distinctions between NB and WB modes were made using colours, with a grey colour showing results for NB (12.5 kHz) and black for WB (100 kHz).

Analysis of the results leads to the following conclusions:

Conclusion No. 1: as in the case of studies discussed in Sect. 7.2.1 for the EM immunity measurements, the narrowband mode (NB) also seems to clearly dominate over wideband (WB) in terms of immunity to extreme multipath propagation conditions. The results obtained for this mode indicate that the PER values are 2.2 times down to 3.4 times lower than for the WB mode, respectively, for the OQPSK and GMSK modulations, averaging within a given channel width (NB/WB).

Conclusion No. 2: in the dynamically changing radio channel obtained during 'dynamic' measurements (corresponding to the mobile channel), the system did not reach PER = 0% in any of the eight tested OMs, remaining at the level of a few percent also in OM_1 mode (0.625 kbps). On the other hand, the system did not lose connectivity in any of the OMs, which was maintained during the test even for OM_8 (100 kb/s), though at the expense of losing approx. 90% of the packets. Meanwhile, in 'static' measurements (a case typical of fixed installations, such as sensors, meters, gauges, etc.), communication in the lowest data rate (OM_1) mode proved to be completely resistant to the multipath propagation. On the other hand, when working in the modes providing the highest data rates (OM_7, OM_8), connectivity was lost completely for more than a half of the stirrer positions, while in the remaining ones, PER topped at values of several dozen percent.

Conclusion No. 3: in spite of the differences in results observed for individual stirrer positions, the average values of PER readings do not differ significantly between 'dynamic' and 'static' modes. In both cases, the total range of PER variation was ~80% in response to the change in R_b from 0.625 to 100 kb/s. The conclusion for both modes is the observation that for OM_7 (50 kbps) and OM_8 (100 kbps), successful transmission may encounter serious problems in conditions of strong multipath propagation. In order to obtain PER below 10%, it is recommended to use the OM_1 (0.625 kb/s) and OM_2 (1.25 kb/s) modes, corresponding to the {OQPSK; NB} configuration.

7.2.3 Summary

The results presented in this chapter allow to formulate a general recommendation regarding the use of specific operational modes in the conditions of strong interferences and multipath propagation.

This recommendation comes down to the statement that the suggested modulation in the Weightless system is OQPSK. Under both types of conditions, it allowed transmission with low PER values due to the spectrum spreading used along with this modulation.

Under favourable propagation conditions, it is recommended to use the wideband variant (WB) with GMSK modulation, as the OQPSK modulation in the NB variant requires a higher CNR level while offering comparable R_b rates.

Regarding the system immunity to multipath propagation, PER response dynamics ranges from 0% to 80%, irrespective of the radio channel type generated in the reverberation chamber, i.e. either stationary or mobile. These are represented by the following measurement modes: 'static' or 'continuous'. However, while in the stationary channel the operating mode OM_1 shows an absolute multipath immunity with PER = 0%, in the mobile channel conditions, the lowest observed PER is approx. 10%.

In the stationary channel, PER averaged over the 12 stirrer positions is approximately 10% lower than in the dynamic channel, for the corresponding OM. Nevertheless, in the stationary channel, a total transmission disruption may occur in the WB GMSK configuration, whereas in the dynamic mode, despite the achievable PER of 90%, the communication was never lost in any of the eight operational modes.

Wrapping up, it can be concluded that the Weightless(-P) system has been well-adapted to long-distance communication in IoT applications, providing a wide range of operational modes. Using these operational modes, it is possible to shape coverage depending on the propagation conditions (expressed by the level of interference/noise and multipath propagation) or application priorities (expressed in the transmission rate and its reliability). The dynamic range of the CNR, within which the system is capable of operating, is 22 dB (as seen in Fig. 7.4), which should ensure high coverage scalability.

7.3 SigFox System

Before discussing the system performance, it should be recalled that according to the system description in Sect. 5.3, the transmission of a single packet is actually done by transmitting the same packet three times on different, random physical channels (Fig. 5.15). This eliminates the need for multiple repetitions of the PER test series at given settings of the hardware and emulating chambers (anechoic and reverberation) to achieve measurement stability, since this mode is naturally enforced by the SigFox communication protocol. The randomness of the three physical channels selection for transmission of a given packet therefore makes *PER* obtained for the 100-packet test sequence correspond to the tests carried out for three independent sequences. The investigation presented in this chapter covers only the UL.

7.3.1 Immunity to Noise and Interference

Due to the functional simplicity of the system, the influence of the controlled AWGN noise is generated by a Tektronix AWG 70002A signal generator, with $BT = 0.3$ and $BW = 250$ kHz, thus completely covering the total 192 kHz-wide bandwidth, designated for SigFox.

The analysis of results presented in Fig. 7.6 leads to the following conclusions:

Conclusion No. 1: the operational assumptions for the system are confirmed as to the order of CNR values (about 7 dB as given in Sect. 5.3), for which the system is able to operate in noisy environments. The positive CNR values for which

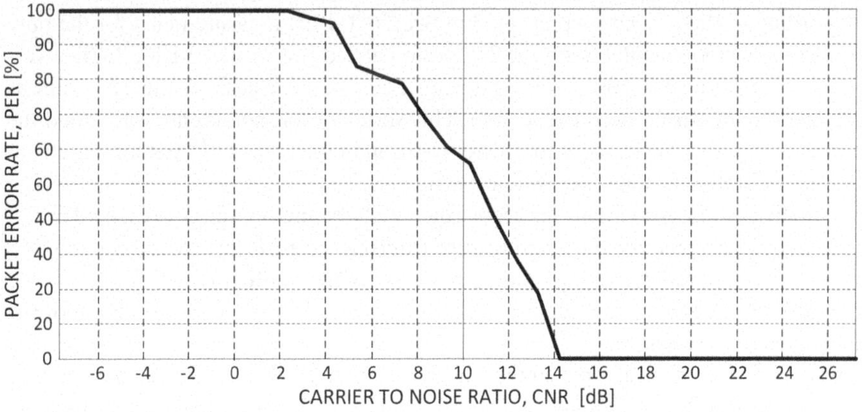

Fig. 7.6 Measurement results of Packet Error Rates for SigFox under controlled interference conditions

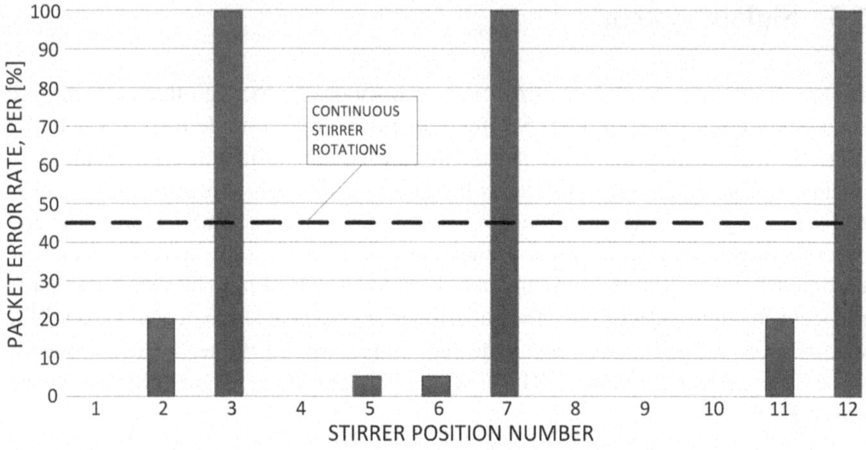

Fig. 7.7 PER measurement results for SigFox in the reverberation chamber

PER<100% result from the use of DBPSK modulation, disregarding the spectrum spreading, which, however, is conducive to maintaining the assumed simplicity of the modems construction.

Conclusion No. 2: unlike the previously analysed two systems, i.e. LoRa and Weightless(-P), for which the dynamic range ΔCNR was approx. 20–22 dB, in SigFox it is limited to approx. 12 dB and is included in range from CNR_{min} = 2 dB to CNR_{max} = 14 dB.

7.3.2 Immunity to the Multipath Propagation

According to the procedure presented in Sect. 6.3, PER measurements for the only operational mode available on the UL were carried out in the WUST ECL reverberation chamber for 100-packet test sequences, one for each of the 12 positions (changed with stirrer steps every 30°). This series of measurements was called the 'static mode' to express the static stirrer position in the course of PER measurement during the transmission of a given test sequence.

Analogical measurements, in accordance with the methodology presented in figure, were also made for the rotating stirrer during the PER test ('dynamic mode'), causing a constantly changing impulse response of the multi-way radio channel, due to continuous changes in boundary conditions for reflections and diffraction of the EM wave in the chamber.

The analysis of results presented Fig. 7.7 leads to the following conclusions:

Conclusion No. 1: in case of 25% of items (i.e. 3 out of 12), stirrers in static mode SigFox showed total transmission inability (PER = 100%). In case of 30% of positions, the negative impact of the multipath transmission led to the loss of packets at the level not exceeding 20%, while in case of other items, the system did not show any susceptibility to the extreme multipath propagation.

Conclusion No. 2: measurements in the dynamic mode reveal an evident decrease in system performance to PER = 46%. However, it exceeds the average PER value (approx. 30%) obtained in the 'static mode' but still ensures successful delivery of approximately every second packet which, in the case of sensor traffic, may be a value acceptable in a number of cases.

7.3.3 Summary

The SigFox system has been developed to provide maximum design simplicity of transceiver devices, based on noise and interference conditions on an ultralow channel width that ensures a relatively high level of the signal spectral density and reduced noise and/or interference power due to such a narrow spectral footprint. The system is characterized by a smaller dynamic range compared to LoRa or Weightless(-P), but the effectiveness of packet delivery in high traffic conditions remains at 75% under static conditions and 55% under dynamic conditions.

Measurements carried out in noise interference and multipath conditions indicate the system good suitability for the machine-type traffic (MTC), especially due to the lack of mechanisms boosting transmission efficiency (such as the spectrum spreading) present in the previously analysed systems.

Chapter 8
The Cellular Internet of Things Systems (CIoT)

The operational environment of mobile systems, led by the 3GPP organization, aware of the market potential of the Internet of Things era, also took steps towards preparing a system offer to handle the machine-type traffic (MTC), which was reflected in defining the mass communication scenario mMTC (see Sect. 1.1) for 5G systems. The first move in this direction was observed in Release 88 (Rel88, Release 88) and subsequent releases aimed at facilitating and improving implementation of M2M transmission in LTE cellular networks, according to the list contained in Table 8.1. Basic changes occurred within the user terminal (UE), for which – starting from Rel8 – several new versions were created, each being better and better adapted to the specific requirements of the Internet of Things devices [44, 119–123]. Technically, the optimization steps have touched on such aspects as:

- The transmission channel width
- The transmission rate
- The duplex mode
- The number of antennas
- Energy-saving modes (among others, by falling asleep and limiting the duty cycle)

In particular, the last three aspects have led to significant simplifications of the receiver radio chain, due to which each subsequent category resulted in reducing the radio modem complexity, as shown in the 'modem complexity' row in Table 8.1. The first steps towards simplifying EU complexity were implemented in the Category 4 (Cat 1, Rel8) equipment, though the most significant changes took place from Cat. 0, defined in Version 12 (Rel 12) for the user equipment in LTE-M(TC) subsystem dedicated to devices handling specific M2M communication applications. This low-cost category defines a set of reduced requirements for less complex and less expensive equipment:

- Enabling half-duplex operation in FDD mode, which allows to avoid duplex filter in the device.

K. Staniec, *Radio Interfaces in the Internet of Things Systems*,
https://doi.org/10.1007/978-3-030-44846-2_8

Table 8.1 List of EU categories in the next LTE 3GPP releases

Version no →	Rel8	Rel8	Rel12	Rel13	Rel13
Name of user terminal equipment category (UE) →	**Cat. 4**	**Cat. 1**	**Cat. 0**	**Cat-M1** (eMTC)	**Cat-NB1** (NB-IoT)
Max. DL rate	150 Mb/s	10 Mb/s	1 Mb/s	200 kb/s	200 kb/s
Max. UL rate	50 Mb/s	5 Mb/s	1 Mb/s	200 kb/s	144 kb/s
Number of antennas	2	2	1	1	1
Duplex mode	Full	Full	Half	Half	Half
BW in UE	1.8–20 MHz	1.8–20 MHz	1.8–20 MHz	1.08 MHz	180 kHz
UE transmitter power	23 dBm	23 dBm	23 dBm	20 dBm	23 dBm
Modem complexity (Referenced to Rel8 Cat. 4)	**100%**	**80%**	**40%**	**20%**	**<15%**

- The receive band has been reduced to a maximum of 1.08 MHz or up to 6 resource blocks (RB).
- LTE-M devices have a single receive path (instead of a double one used for reception diversity).
- Adaptation to low data rates, up to 1 Mb/s, which also entails savings on memory and computing power.
- Limitation of the transmitter radiated power to 23 dBm, which allows it to be integrated with the power amplifier on a common chip.
- Limiting the modem complexity to approximately 40% of the complexity of Cat. 4 devices.

Release 13, in turn, resulted in the creation of two further solutions dedicated to the needs of the machine-type traffic, which are enhanced MTC (eMTC), on the basis of which LTE-M(TC) is developed, and narrowband IoT (NB-IoT), along with their corresponding EU categories: Cat-M1 and Cat-NB1, presented together in Table 8.1. This list allows to relate the radio modem complexity to the devices used in Cat. 4 (as a reference category) optimized for wideband traffic as a priority service, with no mechanisms supporting energy efficiency. The compilation of requirements (with particular emphasis on the 'ultralow complexity') of the mobile Internet of Things system is presented in 3GPP TR 45.820 [45], including a detailed discussion of three system proposals, which will be discussed in more detail later in this chapter, namely:

- EC-GSM M(TC) – *Extended Coverage* GSM M(TC)
- LTE-M(TC)
- NB-IoT

It should be noted that the document [45], in addition to the three documents mentioned above, mentions a greater number of potential systems, but these have been omitted in this study, due to the lack of interest on the part of national regulators, operators and equipment manufacturers. These alternative solutions include narrowband GSM (N-GSM), being a part – along with EC-GSM M(TC) – of a shared group of systems evolved from GSM, as well as other systems: narrowband-OFDMA (NB-OFDMA) and ultra-narrowband cooperative-UNB (C-UNB), which, along with NB-IoT, are together a part of the narrowband M2M (NB M2M) group.

8.1 Functional Assumptions for CIoT Systems

Due to the fact that the concept of IoT systems is a part of the development trend heading to the fifth generation of mobile systems (5G), regardless of the system proposal, each one of these (i.e. EC-GSM M(TC), LTE-M(TC), NB-IoT) is aimed at meeting the requirements described in [45] which all CIoT systems must satisfy (see Sect. 1.1) and, in particular, the criteria for the mMTC scenario, including:

- The boundary coverage (resulting from the target *MCL* value of 164 dB), delay and battery life (3GPP TR 38.913 [124])
- Capacity (ITU-R M.2412 [125], ITU-R M.2410 [126]) and the number of handled devices (ITU-R TR 45.820 [45], ITU-R TR.36.888 [44])

These requirements, compiled for two expected range configurations, A, less demanding, and B, more demanding, are presented in Table 8.2.

In order to reflect the specific character of tasks set for the mMTC family of systems, the mobile autonomous reporting (MAR) model was adopted for the IoT systems intended for the MTS traffic operation [45] defined by parameters presented in Table 8.3. In addition to those, a 99% reporting effectiveness (measured by the number of correctly delivered reports to the base station) is assumed, the boundary *MCL* of 164 dB, 20-byte message size at application layer, 65 bytes of higher layers protocol overhead and a total of 15 bytes of the SNDCP protocol and lower layer protocols overhead – up to the MAC layer (i.e., LLC/RLC/MAC/CRC) – which gives a total of approximately 80 bits of overhead in the MAC layer. The expected total delay in UL delivery is about 10 s for the 'in-band' scenario and about 6.6 s for the 'stand-alone' scenario. This delay is influenced by the following processes that MS must pass in order to obtain the right to transmit:

- Synchronization
- Acquisition of the Master Information Block, contained in the physical broadcast channel
- Multiaccess procedure (messages 1–4), including a waiting time between reception of the random access preamble and a Random Access Response (RAR) from the BS when it is transmitting in downlink
- Permission for UL transmission and duration of data transmission

As for the density of the IoT-enabled devices, in [45] (Annex E.1) and [44] (Annex A), at least 52,547 devices are assumed per a base station sector, which corresponds to 40 IoT devices per household, with the provision (included in [44], based on London and Tokyo) of building densities at 1517 homes per square

Table 8.2 Functional and configuration requirements for mMTC systems ([125])

Parameter	Parameter	
	Configuration A	Configuration B
Inter-site Distance (ISD)	500 m	1732 m
The minimum baud rate at MCL = 164 dB	160 b/s	
Density of mobile MS stations	≥ 1 million devices/ km^2	
Service reliability	≥ 99%	
Delay	≤ 10 s	
Telecommunications traffic model	UL: PDU message of size 32 bytes (in the 2nd layer of ISO/OSI models), once every 2 h, with Poisson distribution DL: Confirmation of size 20 or 65 bytes	
Carrier frequency	700 MHz	
Channel model	Urban macro-mMTC	
Device locations	80% indoor, 20% outdoor	
Building attenuations	20%, strong attenuation; 80%, low attenuation	
Mobility	0–3 km/h	
Base station antenna EIRP	46 dBm/10 MHz, 49 dBm/20 MHz	
Base station antenna configuration	Up to 64 Tx/Rx	
MS mobile station antenna configuration	Up to 2 Tx/Rx	
Reference suspension height of BS (MS) antenna	25 m (1.5 m)	
Single element BS (MS) antenna gain	8 dBi (0 dBi)	
BS noise factor (NF)	According to [125]: 5 dB	
	According to [124]: 5 dB	
	According to [45]: 3 dB	
Mobile station/terminal equipment EIRP	23 dBm	
Mobile station (MS) noise factor (NF)	Acc. to [125]: 7 dB	
	Acc. to [124]: 9 dB	
	Acc. to [45]: 5 dB	
Channel width	≤ 10 MHz	≤ 50 MHz
Multipath propagation models	EPA A, ETU A ([130], [131])	

kilometre and a base station coverage of 1732 m. System simulations performed in [127] indicate even the capability of a single base station to handle up to 72,000 devices.

One of the effective mechanisms in systems dedicated to the MTC traffic is the use of repetitions (repeats) being a form of protection coding leading to a linear processing gain (e.g. 8 repetitions result in $G_p = 9$ dB). This linearity, however, takes place only with the assumption of a long period and a relatively wide frequency range for which the channel estimation is performed. Meeting this condition causes considerable problems, especially in low SNIR conditions (e.g. at distances

Table 8.3 MAR traffic model [45]

Parameter	Value
Statistical distribution of message size at the application layer	The Pareto distribution with the shape factor $\alpha = 2.5$ and the minimum message size at the application layer equal to 20 B and the maximum value of 200 B, i.e. for larger messages, the 200 B size is assumed
Reporting period	The statistical distribution of various reporting periods is distributed as follows:
	1 day: 40%
	2 h: 40%
	1 h: 15%
	30 min: 5%
UL traffic sector load	~ 6.81 reports/second/sector
Percentage distribution depending on the type of traffic (UL)	Data reporting: 80%
	Control commands: 20%

requiring a gain of more than ~12 dB relative to LTE [128]). This is achieved by estimating the channel for multiple subframes and physical resource blocks simultaneously; the efficacy of which and especially for LTE-M has been demonstrated in [44, 129–131]. In the course of standardization of such systems as LTE-M and NB-IoT, it was assumed that the group of terminals (referred to as UE or user equipment) requiring the highest attention are devices located at extremely large distances, with *MCL* close to 164 dB, or located deep inside or underneath buildings (basements, tunnels). Thus, for the modelling of the multipath propagation effects and the Doppler effect on narrowband IoT systems, the following mobile channels were adopted: Extended Pedestrian A model (EPA) and Extended Typical Urban (ETU) model, both with Doppler shift equal to 1 Hz, suitable for both the user equipment (acc. to [132]) and the base station (acc. to [133]). Since both of these channels are, by definition, slow to vary in time and frequency, it is possible to perform a broad channel estimation in both these domains. The fast-varying channel would force to significantly reduce the number of subframes and PRBs covered by a single estimation.

8.2 Energy Efficiency in CIoT Systems

Achieving the second priority goal in CIoT systems, namely, the appropriate capacity (see Sect. 8.1), has become possible mainly due to the implementation of two modes leading to energy savings in UE terminals. However, due to mutual complementarity, each of them can be implemented to maximize energy efficiency, combined with the other or alone. The modes are as follows:

- Power saving mode (PSM) that appeared in 3GPP Release 12 and was defined in [134] (Chapter 4.5.4) and [135] (Chapter 5.3.11)

- Extended discontinuous reception (eDRX), defined in [133][134]

The inclusion of one of these modes (or both together) is by means of a request sent to the network by a UE during the registration or during the implementation of the Tracking Area Update procedure [134], [136].

8.2.1 *Power Saving Mode (PSM)*

A temporary shutdown of the user equipment (UE) is undoubtedly the most preferable way to extend its operation on batteries; however, in order to resume its operation in the network, the UE must reregister in it after switching back to active state. The registration procedure, although it absorbs little energy in a single instance, in case of multiple registrations can significantly contribute to the shortening of the power source lifetime. Meanwhile, in the PSM mode, designed for IoT systems, the terminal remains registered in the network for the entire period of inactivity. The operational procedure of this mechanism consists of the following cycle (also shown in Fig. 8.1):

- The UE may report to the network the desire to enter the PSM mode each time either the TAU or the Routing Area Update procedure is under way. Then, a request to assign time stamps is issued: the 3324 Active Timer stamp and the TAU stamp that is T3412. The assignment of stamps –with values either requested by the UE or imposed by the network itself – depends on the configuration in the local MME/SGSN.
- If the network supports the PSM mode and accepts the EU request, it grants it the T3324 and/or T3412 stamp.
- After sending data from the UE to the network (in the UL), the terminal switches to the idle state by listening for the time of T3324 (given by the AT marker) possible paging from the network side.
- Once the time specified by the AT marker has expired, the device goes into a deep sleep state, in which it does not monitor paging from the network and disables functions which are not critical for waking up.

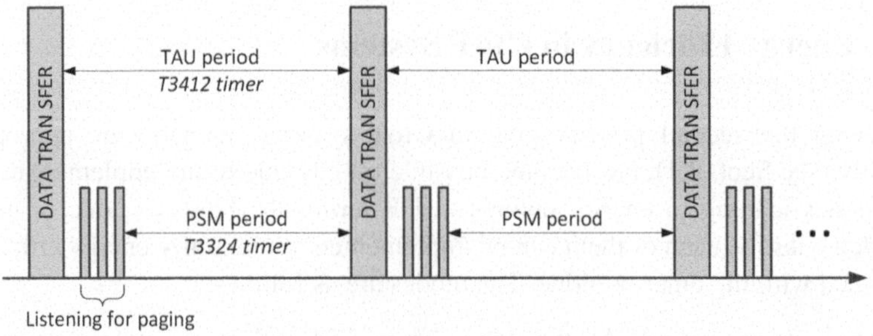

Fig. 8.1 Presentation of the TAU and PSM cycle presentation

The actual energy saving results from the difference between the values of both tags (i.e. T3412-T3324), and although the source documents do not recommend any specific values, it is suggested in [137] that in order to reach a compromise between the duration of the UE in the sleep mode and the effectiveness of communication with the UE on the network part, proportions between the lengths of both markers be given by the formula (8.1). Moreover, it is proposed to adopt a minimum length T3324 (AT) equal to 16 s (at the maximum defined in [140] to be 186 min), while the maximum for T3412 (TAU) equal to 4 h (at the maximum defined in [140] equal to 413 days).

$$\frac{T_{TAU} - T_{AT}}{T_{TAU}} = \frac{T3412 - T324}{T3412} > 90\% \quad (8.1)$$

8.2.2 Extended Discontinuous Reception Mode (eDRX)

The discontinuous mode consists of regularly switching off the device listening and getting into the sleep mode. Due to the fact that this shutdown is designed for two different UE operating phases, discontinuous reception can be divided into DRX in the idle state, i.e. Idle DRX (I-eDRX), and the connected DRX, i.e. C-eDRX, as illustrated in Fig. 8.2.

I-eDRX [138] (Sect. 7.1) is equivalent to the paging cycle, which is based on periodic awakening of the UE to monitor the PDCCH channel, specifically the P-RNTI identifier that informs whether terminal paging is at all included in this channel. When the information about the presence of paging is detected, the UE demodulates the Paging Channel (PCH) to check if it is one of the recipients. When

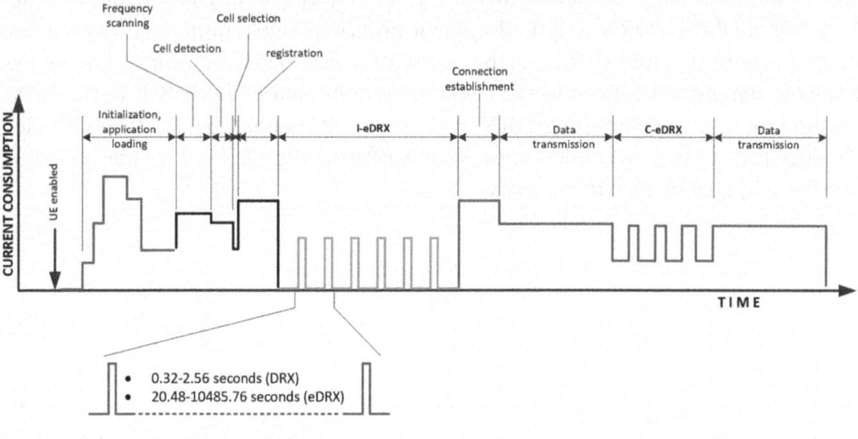

Fig. 8.2 Electric current consumption phases in the UE with indication of eDRX cycles, according to [134]

Table 8.4 The C-eDRX mode parameters

Parameter name	Description
DRX cycle (short/long)	Duration of full 'ON' + 'OFF' cycle, different for the short mode (ShortDrxCycle) and the long mode (LongDrxCycle)
onDurationTimer	Duration of the 'ON' period in the C-eDRX cycle
Drx-inactivity timer	Specifies the length of the UE stay in the 'ON' state after receiving the PDCCH
shortDRXCycle	The number of subframes of a single 'ON' + 'OFF' cycle of the short DRX cycle
drxShortCycleTimer	The number of consecutive subframes intended for the implementation of a short DRX cycle after the 'drx-inactivity timer' has expired

the listening and the possible reception processes are complete, the UE enters the I-eDRX state for the time specified in the second System Information Block (SIB2) contained in the Main Information Block (MIB). Effective return to the listening requires precise synchronization between the base station clock BS and the UE terminal; therefore, the original DRX solution used in LTE [134] assumes spacing between periods of listening up to 2.56 s with the System Frame Number (SFN) as a base for synchronization, the length of which is 10.24 s (i.e. 1024 frames). This period, however, is too short for the machine-type communication (MTC); hence, the Hyper Frame Number (HFN) being 1024 times longer than SFN was assumed as the basis for synchronization here, with the duration of about 175 h (10,485 s). The synchronization between the UE and BS is performed using the MIB, whose first information element (IE) is the SFN number. If the base station supports the extended DRX mode (eDRX), information about the current HFN value is transferred in a special IE within SIB1.

C-eDRX [139] (Sect. 5.7) is a mechanism with a lower efficiency in terms of the energetic efficiency than I-eDRX, because – as opposed to the inactive mode – the low-energy mode consists only in turning off the receiver module (i.e. transiting into the inactive state), rather than switching to the sleep mode. The network informs the UE about the C-eDRX settings during the procedure of obtaining radio resources, Radio Resource Control (RRC), in the course of which the UE is informed about the values of parameters related to the DRX cycle presented in Table 8.4. In the basic configuration, a so-called long DRX cycle is set up (defined by the 'DRX Cycle' parameter, as in Fig. 8.3) filled with periods of listening (defined by the 'onDurationTimer' parameter) and inactivity.

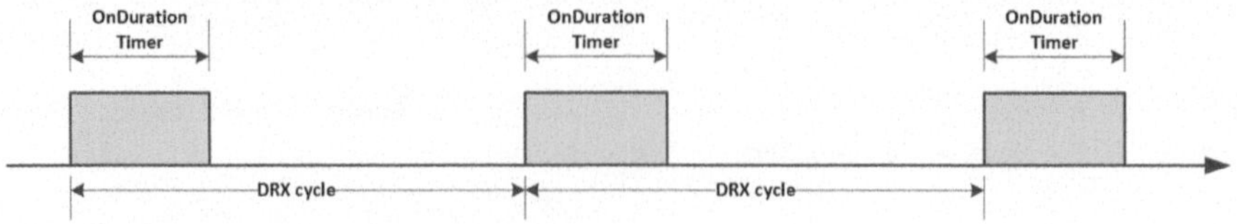

Fig. 8.3 Basic configuration of the long DRX cycle

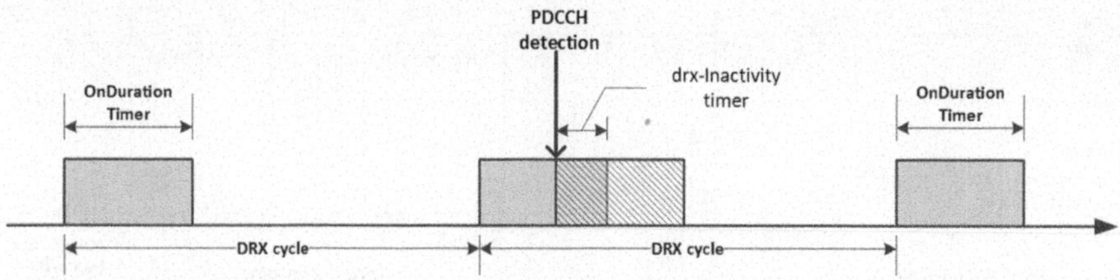

Fig. 8.4 A transition to the long DRX cycle after receiving the PDCCH

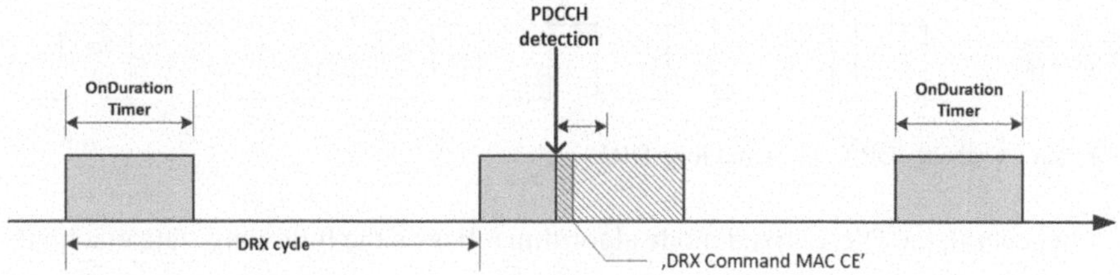

Fig. 8.5 An accelerated transition to a long DRX cycle after receiving the PDCCH

After receiving the PDCCH channel during the listening period, the UE decodes it until the command 'DRX command MAC CE' has been received, which heralds the end of transmissions relevant for a given UE and at the same time orders a transition to the long DRX cycle state or – if activated in the cell – first to the short DRX cycle (the length of which is defined by the 'shortDRX-Cycle' parameter), after which a long DRX cycle will begin. If a long DRX cycle is set in the cell, two scenarios are possible:

- From the moment of PDCCH reception, the 'drx-Inactivity Timer' is activated, and once finished, UE switches to the inactive state, as in Fig. 8.4.
- After receiving the PDCCH, UE may receive the 'DRX MAC CE' command forcing immediate transition into the inactive state, which results in an instant (accelerated) zeroing of the following timers: 'drx-Inactivity Timer' and 'onDurationTimer' (Fig. 8.5).

The operational principle of a mixed configuration, i.e. one involving both the long and the short DRX cycle, is presented in Fig. 8.6, for exemplary timer setting values:

- drx-InactivityTimer = 3
- onDurationTimer (Long/Short) = 2
- ShortDRXCycle = 5
- LongDRXCycle = 11
- drxStartOffset = 0
- drxShortCycleTimer = 3

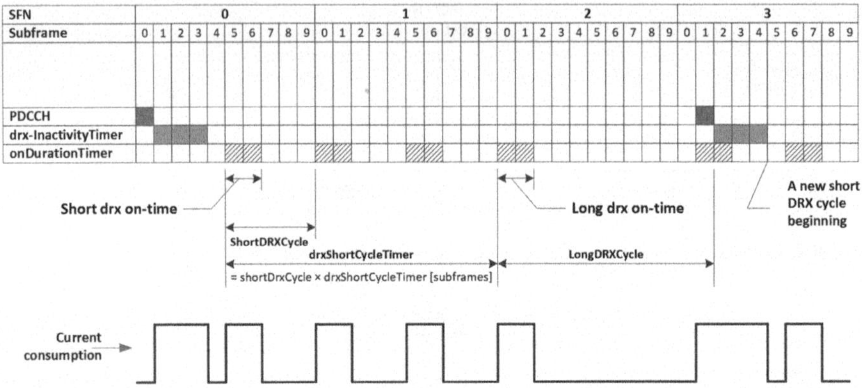

Fig. 8.6 A mixed eDRX: short and long DRX cycles

In general, the DRX mixed mode algorithm relies on the following state machine:

1. Once the C-eDRX mode is activated, the PDCCH is received.
2. The 'drx-Inactivity Timer' is activated, during which the UE remains in the listening state until the time has reset.
3. Once the set time has elapsed from expiry of the 'drx-Inactivity Timer', a short DRX cycle with the total duration of 'drxShortCycleTimer' is activated, in which the listening periods (each with 'onDurationTimer' duration) are separated by periods of inactivity.
4. After the short DRX cycle has expired and the PDCCH has not been detected in this period, the long DRX cycle with a longer inactivity period is activated in a single cycle. Within it, assuming sufficiently long periods of inactivity, one can activate the short DRX cycle.
5. If, in the course of any wake period (i.e. active listening), the PDCCH channel is detected, the algorithm returns to pt. 2.

8.3 NB-IoT: Narrowband IoT

As mentioned before, NB-IoT appeared in Release 13 (Rel13) 3GPP [45] and in ETSI document [141] with the following set of functional assumptions:

- Extended indoor coverage: by increasing the link energy budget by 20 dB compared to GPRS receivers (for UL connection, whose *MCL* = 144 dB), which corresponds to NB-IoT attenuation dynamics at the level of 164 dB. For this value, the minimum transmission rate at the application layer for UL and DL should be at least 160 b/s. Two mechanisms in particular are used to achieve these assumptions:

– Repetitions, used both in UL channels (e.g. NPRACH, NPUSCH) as well as DL channels (e.g. NPDCCH, NPDSCH), as a form of redundant coding consisting in multiple repetitions of the same message from higher layers in a given channel (a transport block) in subsequent subframes (the time structure of the DL transmission frame shown in Fig. 8.9). Their number, N_{rep}, is determined by eNB according to the signal strength detected and reported by the UE to the base station. Based on this information, the eNB categorizes the UE equipment into various categories called (CC) coverage classes. These classes determine N_{rep} – with the highest values assigned to the EUs experiencing the greatest pathloss or smaller values assigned to the EUs located in more favourable propagation conditions, i.e. ones with lower MCLs. 16 levels are available for the downlink: $N_{rep} = 2^i$ ($i = 0,1, 2, \ldots, 15$) which yields up to 2048 repetitions (Table 16.4.1.3–2 in [142]), while for the uplink, there are 8 levels $N_{rep} = 2^i$ (i = 0,1, 2, ..., 7), i.e. up to 128 repetitions (Table 16.5.1.1–3 in [142]).
– Scalable width of the allocated multi-tone transmission channel in UL, in the range from 3.75 kHz (which corresponds to ¼ of the subcarrier width) to 180 kHz (12 subcarriers).

• Reduced complexity of client devices transceiver systems.
• Improved energetic efficiency: allowing for power supply from a single 5 Wh battery of up to 10 years under MCL = 164 dB.
• Delay: up to approximately 10 s with 99% connection efficiency.

For NB-IoT, there are also three implementation scenarios providing operators with flexibility during system implementation, depending on the availability of spectrum resources. These scenarios are shown schematically in Fig. 8.7 and their actual implementation in the form of measured spectra in Fig. 8.8.

In the ‘stand-alone’ scenario NB-IoT occupies one or more independent GSM channels, which supports the process of GSM refarming.

In the ‘in-band’ scenario, one or more physical blocks of PRB resources from the LTE channel are reserved for NB-IoT transmission. Within this reservation, however, NB-IoT signals cannot use frequency-time resource components used in LTE for control or reference purposes. On the other hand, the radiated power EIRP is shared with other LTE resource blocks, although to increase the spectral power

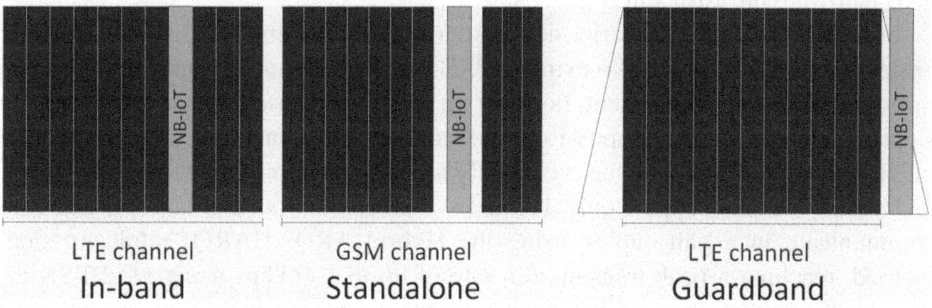

Fig. 8.7 Implementation scenarios for NB-IoT

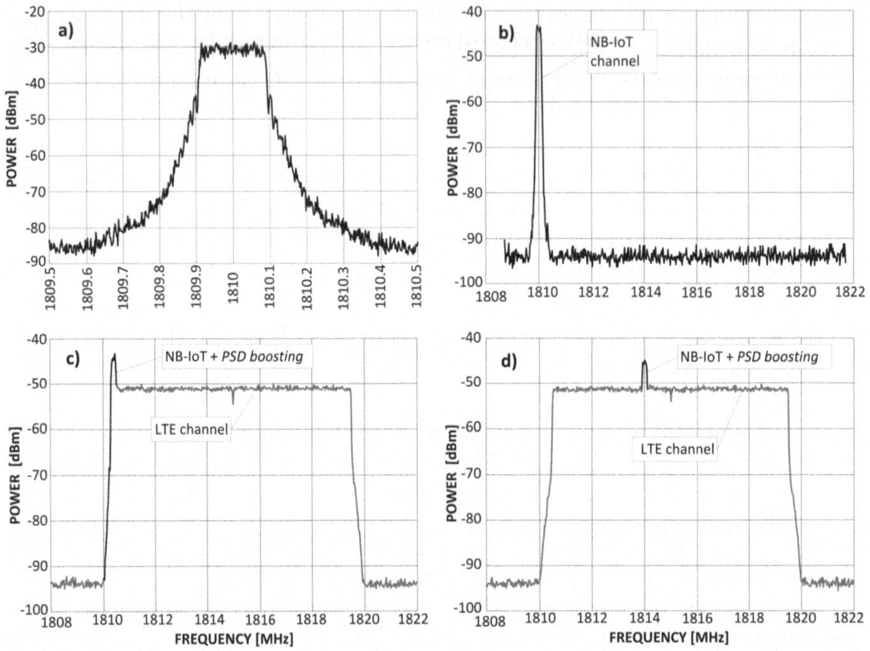

Fig. 8.8 Measured NB-IoT signals spectra: (a) an isolated channel and channels in scenarios:(b) stand-alone; (c) guard band; (d) in-band

density (PSD) of the NB-IoT channel, the mechanism of the so-called PSD boosting is used by providing extra 6 dB to this channel, in comparison with other PRB in the shared LTE channel. The effect of this mechanism is particularly visible in the spectrum of the NB-IoT signal presented in Fig. 8.8, where the resource block dedicated to NB-IoT stands out from the remaining 49 PRBs constituting the LTE channel width of $BW = 10$ MHz, whose individual powers are reduced evenly with respect to the total transmission power by 17 dB (i.e. by 10·log50). This scenario allows for more efficient spectrum management and simple increase of NB-IoT capacity as the number of devices in the network increases. An additional advantage of this arrangement is the fact that both systems, i.e. LTE and NB-IoT, can be operated by the same eNB hardware infrastructure.

In the 'guard band' scenario, new resource blocks defined in the LTE channel protection band next to its two extreme PRBs are used to implement NB-IoT transmissions. Such an arrangement, however, may entail the use of filters with steeper slopes, especially due to the possibility of using the abovementioned PSD boosting in NB-IoT channels. This issue is currently at the stage of intensive research (. 8.9).

NB-IoT stations support only 180 kHz channels on both DL and UL and can communicate in a half-duplex using the Hybrid ARQ (HARQ) retransmission method, reaching a peak transmission rate of up to 100 kbps using π/2-BPSK or π/4-QPSK modulation in the SISO mode.

	subframe number									
	0	1	2	3	4	5	6	7	8	9
even frame	NPDCCH	NPDCCH or NPDSCH	NPDCCH or NPDSCH	NPDCCH or NPDSCH	NPDCCH or NPDSCH	NPSS	NPDCCH or NPDSCH	NPDCCH or NPDSCH	NPDCCH or NPDSCH	NSSS

	subframe number									
	0	1	2	3	4	5	6	7	8	9
odd frame	NPDCCH	NPDCCH or NPDSCH	NPDCCH or NPDSCH	NPDCCH or NPDSCH	NPDCCH or NPDSCH	NPSS	NPDCCH or NPDSCH	NPDCCH or NPDSCH	NPDCCH or NPDSCH	NPDCCH or NPDSCH

Fig. 8.9 Time-frequency allocation of channels in the DL link

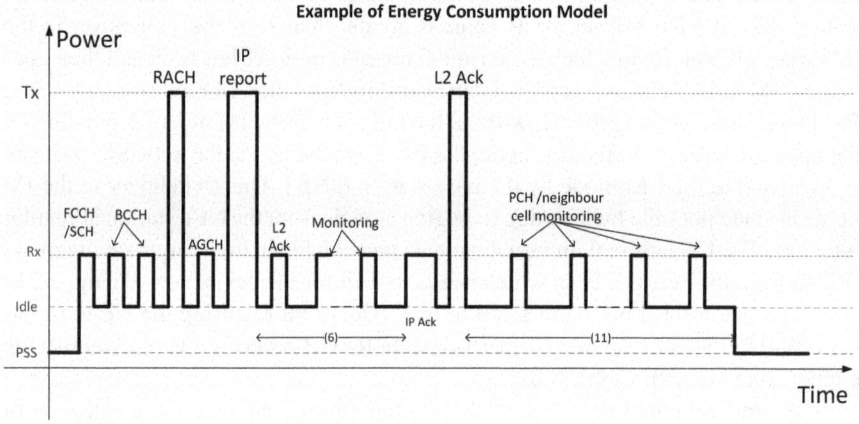

Fig. 8.10 Examples of events affecting the energy consumption during IP packet transmission. (Based on Fig. 5.1–4 in [45])

Simultaneous application of all the above-described mechanisms allowed to reduce the level of transceiver complexity below 15% in relation to the LT8 Rel8 model (Cat. 4), as shown in Table 8.1. An additional mechanism facilitating the energetic efficiency of the NB-IoT UE was the assignment of different energy levels to equipment, depending on the transmission/reception phase and confirmations, as shown, for example, in Fig. 8.10 for the IP packet transmission process. Power consumption required to implement the states shown in the diagram is presented in Table 8.5.

Table 8.5 A model of power consumption in NB-IoT devices

MS device mode	Power consumption [mW]
Transmission (CM)	543
Reception (CM)	90
Idle mode (IM)	2.4
Sleep mode (PSM)	0.015

The UE in the NB-IoT system can be in one of three modes:

- Connected Mode (CM)
- Idle Mode (IM)
- Power saving mode (PSM)

In the CM mode, the user terminal receives PDCCH messages and passes PDUs to the MAC layer along with the signalling of the non-access layer of the NAS (Non-Access Stratum), also responsible for the support of mobility and authentication, session management (including establishing, sustaining and ending the connection) as well as higher layer data, via the radio interface. In the absence of UE activity for a predefined period of time, the device can be switched to the IM or PSM mode. The transition to one of these states is associated with the Activity Timer (AT). If AT is not set, or its value is greater than zero, the switchover to the IM mode takes place, in which the terminal remains until AT has reset and then goes to the PSM mode. The user terminal can also shift from the CM mode directly to the PSM mode after zeroing the AT, without having to enter the IM mode, depending on the operator settings. In the IM mode, the EU is reachable via the network, so it can be switched to the CM mode by the base station (eNB). This switchover in the IM mode is made possible by listening to paging signals from the base station at regular intervals. The UE terminal, in turn, becomes unavailable in the energy-saving mode (PSM) from the network level which needs to wait for the device to wake up, either after a predetermined period of sleep or due to an event requiring the UE to report to eNB. After waking up, the UE performs the RACH procedure to re-establish the connection with eNB (Table 8.6).

New physical channels and so-called signals (occupying individual elements of the resource block) were defined in NB-IoT and are presented in Table 8.7: 3 channels and 2 signals in the DL and 2 channels and 1 signal in the UL. It is worth noting that, contrary to LTE, the control channel on the UL was abandoned in NB-IoT. HARQ

Table 8.6 Combinations of resource elements (REs), forming Resource Units (RU) in NB-IoT, acc. to [141]

NPUSCH format	Δf	N_{sc}^{RU}	N_{slots}^{UL}	N_{symb}^{UL}
1	3.75 kHz	1	16	7
	15 kHz	1	16	
		3	8	
		6	4	
		12	2	
2	3.75 kHz	1	4	
	15 kHz	1	4	

Explanations of symbols:
Δf – Distance between the OFDM subcarriers
N_{sc}^{RU} – Number of adjacent subcarriers in one RU
N_{slots}^{UL} – Number of adjacent time slots in one RU
N_{symb}^{UL} – Number of OFDM symbols in a single time slot

Table 8.7 New transport and physical channels in the NB-IoT system

Channel	
DL	Narrowband Physical Downlink Control Channel (NPDCCH) transfers paging information to specific UEs and transfers responses to random RACH access requests, DL/UL transmission assignments and ACK transmission acknowledgements in PUSCH channel. It occupies a specified number of subcarriers in each of the 31 slots of the subframes
	Narrowband Physical Downlink Shared Channel (NPDSCH) is designed to carry data transmission and any other data not allocated to PSCH, PDCCH and PBCH channels
	The Narrowband Physical Broadcast Channel (NPBCH) carries information about the network in general and the configuration specific to the given eNB included in the Main Information Block (MIB). NPBCH is transmitted in the zeroth subframe of each frame, i.e. once every 10 ms. The MIB remains unchanged throughout the Transmission Time Interval (TTI) period of 640 ms
	Narrowband sync signals, primary and secondary, respectively, NPSS/NSSS (Narrowband Primary/Secondary Synchronization Signal), allow the UE to synchronize with the eNB in the time and frequency domain. The NPSS signal is transmitted every 10 ms in the 5th subframe of each frame using its last 11 OFDM symbols. NSSS signals are transmitted in the 9th subframe of each even frame, occupying also the last 11 OFDM symbols. In both NPSS and NSSS signals, pseudorandom Zadoff-Chu sequences are used, due to their autocorrelation properties and constant amplitude
	Narrowband Reference Signal (NRS): It is used as a phase reference in DL channel demodulation. Each antenna port has eight NRS signals transmitted in all subframes on eight disjoint resource elements. Additionally, in the 'in-band' mode in the NB-IoT channel, cell-specific signals, typical for LTE (so-called CRS, Cell-specific Reference Signal), are also transmitted. They are absent in 'stand-alone' and 'guard band' scenarios
UL	Narrowband Physical Uplink Shared Channel (NPUSCH)
	Narrowband Physical Random Access Channel (NPRACH), in the form of a preamble, occupies a set of adjacent 12, 24, 36 or 48 subcarriers located at randomly changed frequencies during each subsequent repetition (at N_{rep} up to 128). The base station uses these preambles to estimate the propagation delay on the UL, necessary to issue a timing advance command to maintain orthogonality in the UL between different users
	Demodulation Reference Signal (DMRS), interleaved between data transfer RE, occupying one to three symbols per slot

confirmations along with the remaining signalling are sent via the NPUSCH channel, and the distinction between data and control information is made by using one of two formats. Format 1 is used for data transmission using turbo codes for error correction. Format 2 is used for transmitting the abovementioned HARQ confirmations for DL data. Also characteristic is the ability to work on the UL using a different number (i.e. 1, 3, 6 or 12) of carriers of width of 15 kHz each and a slot with a length of 0.5 ms. One-tone transmission is also possible in a twofold grid: with a 15 kHz and 3.75 kHz carrier space and with a gap of up to 2 ms in the second case. Each of these slots thus contains 7 OFDM symbols and 48 subcarriers. The UL frame, 10 ms or 40 ms long, thus contains 10 subframes, in which the NPUSCH and NPRACH channels are broadcast. Specifically for NB-IoT, the so-called Resource Unit (RU) was defined as a combination of the number of subcarriers, time slots and OFDM symbols (always 7) in a single time slot. Possible combinations are listed in Table 8.6.

Owing to the NPSCH channel, after decoding the synchronization signals carried in it, i.e. the Primary Synchronization Signal (PSS) and the Secondary Synchronization Signal (SSS), the UE obtains the time-frequency synchronization with eNB, thus becoming capable of reading the System Information (SI) about the cell, including the cell identity, that is broadcast in the NPBCH channel present in the next time slot in the NPSCH channel. There are four physical blocks of PRB resources per a NPBCH channel, one in each even subframe, each PRB containing the same message, thus increasing the processing gain G_p by means of Tx diversity. One of the items contained in the System Information is the Mandatory System Information (MSI), describing the PDCCH message structure described in the 'PDCCH configuration' field. This information is necessary for the UE to choose an appropriate coverage class (CC), described later in this chapter, which facilitates efficient management of available energy resources by a UE (usually battery-powered). In the NPDCCH channel, the base station contains information on the UL and DL transmission assignment, including normal data transmission, RACH response, paging messages and other control messages. The channel occupies multiple subcarriers and slots in which it transmits messages regarding the UL and DL traffic scheduling, coding messages for each UE independently.

For the energy requirements defined in this way, the validation for the long-term operation of the NB-IoT device on a single battery proves true, as shown in Table 8.8. The calculations were made for two packet sizes and reporting intervals, assuming equipment operation on the coverage edge, i.e. for $MCL = 164$ dB. Clearly, reporting seldom, with a frequency of one message per day (still enough, e.g. for the construction of individualized media usage profiles, such as gas, electricity, water, etc.), allows equipment to operate without battery replacement, even for more than 10 years, for both packet sizes: 50 B and 200 B.

As for the radio interface in DL, both the subcarrier grid (measured by the distance between subcarriers equal to 15 kHz) and the multiplicity of the OFDM channel remained unchanged compared to LTE. In UL the option of transmission has been assumed using NPSCH shared channels in multi-tone mode with a channel spacing of 15 kHz (1, 3, 6 and 12 multiples of 15 kHz, as shown in Fig. 8.11, with 12 being supported as a mandatory requirement for the terminal) and using – similarly as in LTE – SC-FDMA multiple access. Single-tone transmission with a 15 kHz channel spacing is also possible (which gives 12 subcarriers in one PRB) and a 3.75 kHz channel spacing (which gives 48 subcarriers in one PRB). It should be noted that in the in-band scenario, the use of a single tone transmission (i.e.,

Table 8.8 A model of a battery lifetime (in years) in NB-IoT devices

	Estimated lifetime with no battery replacement [in years]	
Packet size/reporting interval	Stand-alone mode	Guard band mode
50 B/2 h	2.6	2.4
200 B/2 h	1.2	1.2
50 B/1 day	18.0	16.8
200 B/1 day	11.0	10.5

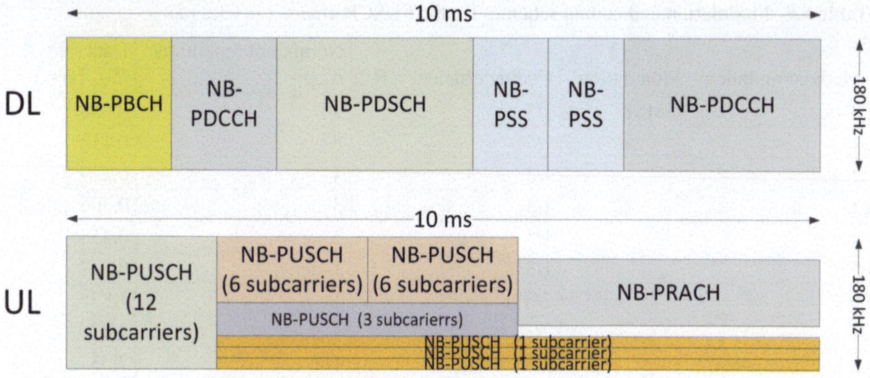

Fig. 8.11 Sample configurations of the PHY frame in the NB-IoT system

using a single subcarrier) will cause interference between NB-IoT and LTE which, however, can be minimized by queuing users with similar SNR requirements using NB-IoT channels and adjacent blocks of PRB LTE resources. The risk of similar disturbances, however, does not exist in the case of using 15 kHz grid between subcarriers, due to the orthogonality of LTE and NB-IoT subcarriers. It is worth emphasizing that both the use of single tone transmission and the previously described mechanism of 'PSD boosting' are aimed at achieving a target energy budget increased by 20 dB relative to the GPRS system. It becomes possible since UE capable of transmitting in the single-tone mode, in addition to the low noise level attributed to the small channel width (i.e. 3.57 kHz or 1, 3, 6, 12 multiples of 15 kHz), provides also a low level of peak-to-average power ratio (PAPR), which allows the power amplifiers (PA) to work with a greater efficiency: a parameter limited by the size of the linearity range.

The NB-IoT system was developed to efficiently allocate the available capacity to many users while occupying minimum bandwidth, with a maximum EIRP of UE devices not exceeding 23 dBm (while, for comparison, GPRS mobile stations had $EIRP_{max} = 33$ dBm). The EIRP of base stations (BS), on the other hand, depends on the type of scenario [143] (Table C.1 i 4.6):

- In the 'in-band' scenario, $EIRP_{max} = 46$ dBm, due to the channel sharing with the LTE system. Since this power has been determined for a reference LTE channel containing 50 PRB, with a width equal to or greater than $BW = 10$ MHz (for narrower LTE channels LTE $EIRP_{max} = 43$ dBm is assumed), a single PRB is allocated: 46 dBm-10log(50) dB or 29 dBm. After the addition of 6 dB due to the use of the abovementioned 'PSD boosting' mechanism, the actual $EIRP_{max} = 35$ dBm per single NB-IoT channel.
- In the 'stand-alone' scenario, due to frequency band sharing with the traditional GSM system, also the principles of EIRP selection are subject to regulations regarding this system; hence $EIRP_{max} = 43$ dBm for each GSM channel with a width of 200 kHz.

Table 8.9 Modulation and coding schemes for the PUSCH channel in Class 1

Modulation index	Modulation	Coding efficiency, R	Number of repetitions, N_{Rep}	Data rate R_{PHY} [kb/s]
0	GMSK	1/3	16	0.058
1		1/3	8	0.115
2		1/3	4	0.229
3		1/3	3	0.306
4		1/3	2	0.458
5		1/3	1	0.917
6		2/3	1	1.83
7		2/3	1	3.67
8		2/3	1	7.33
9		2/3	1	14.7

Table 8.10 Modulation and code schemes for the PUSCH channel in Class 2

Modulation index	Modulation	Coding efficiency, R	Number of repetitions, N_{Rep}	Data rate R_{PHY} [kb/s]
0	π/2-BPSK	1/3	16	0.063
1		1/3	8	0.125
2		1/3	4	0.25
3		1/3	3	0.33
4		1/3	2	0.50
5		1/3	1	1.0
6	π/4-QPSK	1/3	1	2.0
7		2/3	1	4.0
8		2/3	1	8.0
9		2/3	1	16.0
10		2/3	1	32.0
11	π/8-8PSK	2/3	1	48.0

Regarding the modulation-code schemes (MCS), two modulation classes for the uplink have been defined for the terminals, in various combinations with the number of N_{Rep} repetitions and the code rate R, resulting in available PHY data rates R_{PHY}, compiled for both classes in Table 8.9 and Table 8.10 (a list of available downlink MCSs for PDSCH is presented in Table 8.11):

- Class 1. (Class-1): based on GMSK modulation, obligatory for all NB-IoT UEs.
- Class 2. (*Class-2*): based on PSK modulation in the following variants: π/2-BPSK, π/4-QPSK and π/2-8PSK. The class is optional for UE's.

Such a high scalability of transmission modes (MCSs) allows to obtain a significant flexibility in adjusting transmission parameters to the available link budget (mainly determined by route attenuation) and quality requirements (priorities) depending on the specific application. In NB-IoT, the concept of the coverage class

Table 8.11 Modulation and code diagrams for the PDSCH channel

Modulation index	Modulation	Coding efficiency, R	Number of repetitions, N_{Rep}
0	BPSK	1/3	8
1	BPSK	1/3	4
2	BPSK	1/3	2
3	BPSK	1/3	1
4	BPSK	1/2	1
5	BPSK	2/3	1
6	QPSK	1/2	1
7	QPSK	2/3	1
8	8PSK	2/3	1
9	16-QAM	1/2	1
10	16-QAM	2/3	1

Table 8.12 PDCCH channel configurations for different reference coverage classes

Coverage class (CC)	1	2	3	4
Modulation	QPSK	BPSK	BPSK	BPSK
Coding rate, R	2/3	0.444	4/9	0.061
Number of repetitions, N_{Rep}	1	1	2	2

(CC) was introduced, allowing the operator to use different transmission parameters at terminals, depending on the relevant CC. As a result, the UE working under MCL conditions lower than the maximum (i.e. 164 dB) can adjust the transmission length, modulation, number of repetitions, transmit power, etc. in a way that maximizes battery life and allows access to BS with optimal operational parameters, on which the entire network also gains as a result of increasing the total capacity in the cell.

From the UE point of view, a device under good propagation conditions will be able to operate at higher transmission speeds and lower latencies than the UE located in a more challenging propagation situation (larger MCL). These classes are reflected mainly in the Physical Downlink Control Channel (PDCCH), which was defined for four resource configurations (including modulation, coding rate, number of repetitions and arrangement of subframes), one for each of the four reference range classes, as given in Table 8.12 . Apart from these four major classes, the document [45] suggests a distinction between three reference coverage classes, related to GPRS coverage. These CCs have also been accepted by operators as well as hardware and software producers as a good practice:

- Normal CC, corresponding to $MCL = 144$ dB, which is equivalent to the GPRS range limit
- Extended CC, corresponding to MCL increased by 10 dB compared to GPRS (meaning $MCL = 154$ dB)
- Extreme CC, corresponding to the MCL plus a target of 20 dB relative to GPRS (i.e. with $MCL = 164$ dB)

The UE selects the best CC class for PDCCH monitoring, based on the estimated received signal quality in the downlink (DL). A typical procedure for preselecting the correct CC consists of the following steps:

- Step 1: The UE measures the pilot PSS/SSS signals present in the PSCH channel.
- Step 2: The UE retrieves the PDCCH configuration from the system information.
- Step 3: The UE selects the most likely CC candidate, based on the measurement results on the DL.
- Step 4: The UE attempts to decode the PDCCH message corresponding to the preselected CC candidate and then makes a selection:
  - If the predefined success criterion (i.e., a certain number or percentage of successfully decoded PDCCH) has been met, the candidate CC becomes valid for subsequent connections.
  - If the assumed criterion has not been met:
    - If the CC candidate index is lower than the highest (e.g. equal to 4 in Table 8.12), the CC candidate index is incremented by 1 and then UE returns to Step 4.
    - If the CC candidate index is maximal, the UE may start the procedure of selecting a new cell.

The procedure described here can also be performed during a normal UE operation, after selecting the appropriate CC, in order to adapt the current CC to new conditions. It may be advisable, for example, if the pathloss has altered (which may happen dynamically in mobile scenarios). Regular detection of messages carried in the PDCCH channel enables the implementation of the CC selection procedure starting from Step 2.

Table 8.13 presents the entire energy budget of the DL and UL. It can be concluded that in the 'in-band' variant, the data rates above SNDCP (equivalent to the third layer of the ISO/OSI model, located directly above the IP layer and under the LLC layer [144]) achievable in DL are significantly lower than in the 'stand-alone' scenario. This is due to a significant, 8 dB, difference between the two options in the actual base station EIRP. As already mentioned, in the 'in-band' variant, total radiated power is proportionally shared among the remaining PRB resource blocks (one of which is used by NB-IoT), while in 'stand-alone' variant, the total EIRP is dedicated to each of the 200-kilohertz GSM channels, including the NB-IoT channel situated among them.

8.4 LTE-M(TC): LTE-Machine (Type Communication)

The main concept behind the development of the LTE system was to create a solution slightly different from the traditional LTE, while meeting the requirements for CIoT systems (IMT-2020) described in Table 8.2. The works proceeded towards

Table 8.13 Link energy budget for the scenarios: 'in-band' and 'stand-alone'

Scenario	In-band		Stand-alone	
Channel	NPDSCH	NPUSCH	NPDSCH	NPUSCH
Data rate above SNDCP [kb/s]	0.44	0.31	2.73	0.31
Transmitter				
$EIRP_{max}$ [dBm]	46	23	43	23
(1) Actual $EIRP_{max}$ (per PRB) [dBm]	35	23	43	23
Receiver				
(2) Spectral density noise [dBm/Hz]	−174	−174	−174	−174
(3) Receiver noise factor [dB] (acc.To [45])	5	3	5	3
(4) Interference margin [dB]	0	0	0	0
(5) Actual channel width [kHz]	180	2.5	180	2.5
(6) Effective noise power [dBm] = (2) + (3) + (4) + 10log((5))	−116.4	−137	−116.4	−137.0
(7) Required SNIR [dB]	−13.7	−5.8	−5	−5.8
(8) Receiver sensitivity [dBm] = (6) + (7)	−130.1	−142.8	−121.4	−142.8
(9) Receiver processing gain [dB]	0	0	0	0
(10) MCL = (1)–(8) + (9) [dB]	165.1	165.8	164.4	165.8

Table 8.14 Model power consumption in UE LTE-M devices

MS device mode	Power consumption [mW]
Transmission (CM)	575
Reception (CM)	80
Idle mode (IM)	3
Sleep mode (PSM)	0.015

simplifications in both construction and transmission, beginning with LTE Release 8 (Rel8), as shown in Table 8.1. The current shape of LTE-M was accomplished in 2016 in 3GPP TR 45.820 [45]. Similar to NB-IoT, the system was supposed to reach *MCL* higher by 22 dB, compared to traditional LTE (which is 142 dB), and work on a single AA battery (5 Wh) for at least 10 years (with the current consumption shown in Table 8.14), while complying with the LTE channel raster due to the following modifications:

- Reducing the number of DL control channels (PDCCH, PCFICH (Physical Control Format Indicator Channel) and PHICH (Physical Channel HybridARQ Indicator Channel)) to just one: EPDCCH (*Enhanced Physical Downlink Control Channel*). This channel is also called MPDCCH (MTC PDCCH).
- Limiting the maximum channel width to 6 PRBs, i.e. using LTE channels with the smallest possible width of 1.4 MHz (or 1.08 MHz of the actual width).
- Relinquishing MIMO multi-antenna solutions in favour of SISO 1 × 1.
- Decreasing the transmission rate to 1 Mb/s or approx. 200 kb/s.
- Limiting the Transmit Block Size (TBS) to 1000 bits per subframe.

Table 8.15 Maximum number of repetitions for different range configurations in LTE-M

LTE-M channel	Number of repetitions	
	Configuration A	Configuration B
PSS/SSS	1	1
PBCH	1	5
MPDCCH	16	256
PDSCH	32	2048
PUSCH	32	2048
PUCCH	8	32
PRACH	32	128

- Specifying two UE power classes, which allows the use of built-in power amplifiers (integrated with the modem):
  - PC 3: with a maximum EIRP of 23 dBm
  - PC 5: with a maximum EIRP of 20 dBm
- Introducing the mechanism of 'PSD boosting' in DL by 4 dB relative to LTE.

Actual values of *MCL*, quoting after [128–146], range from 155.7 dB for UE class PE 5 to 161 dB for UE PE 3 class, achieving targets in the context of energy budget, based on the use of repetitions and 'PSD boosting', as in NB-IoT (Sect. 8.3). One example of the coverage scaling by means of gain resulting from the use of repetitions is a set of recommendations specified for correct reception of different channels depending on the coverage configuration (A or B, as in Table 8.2) presented in Table 8.15. Although the use of a UE with a higher power class (PC 3) seems to be another natural step towards greater *MCL*, the LTE-M specification contains a variant with a lower EIRP (i.e., PC 5), because research has confirmed that higher power ranges often attract different sort of practical problems, e.g. increased UE costs, regulatory issues related to the EM radiation (and resultant dose limits), increased interference or problems with amplifiers linearity. Similarities to LTE can also be seen in the method of calculating the energy budget for LTE-M shared channels. The budget is based on LTE FDD parameters (Table 5.2.1.2–2 w [44]), with the only difference lying in accepting the SNIR in DL 4 dB higher than in LTE, due to the mentioned 'PSD boosting' mechanism.

An example of transmission in DL has been depicted In Fig. 8.12. At small distances (corresponding, e.g. to Configuration A), it is possible to use the entire

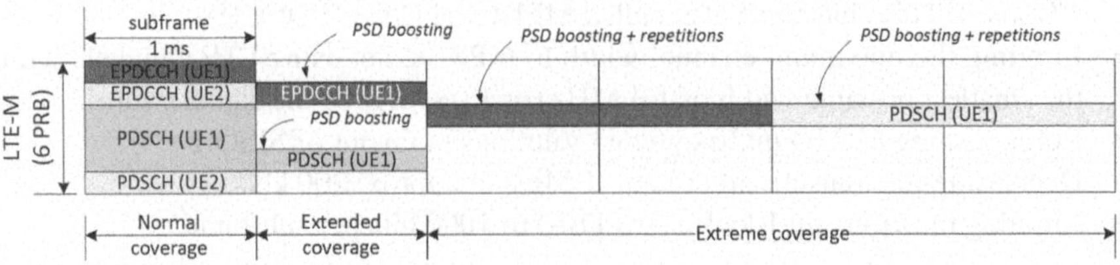

Fig. 8.12 A diagram presenting LTE-M DL transmission in a 1.4 MHz channel [146]

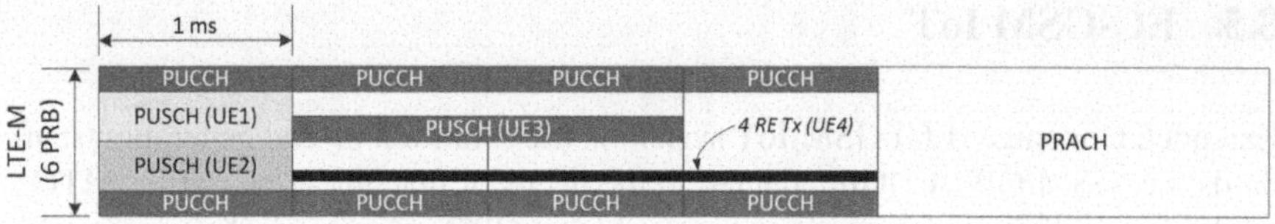

Fig. 8.13 LTE-M transmission for UL connection in the 1.4 MHz channel [146]

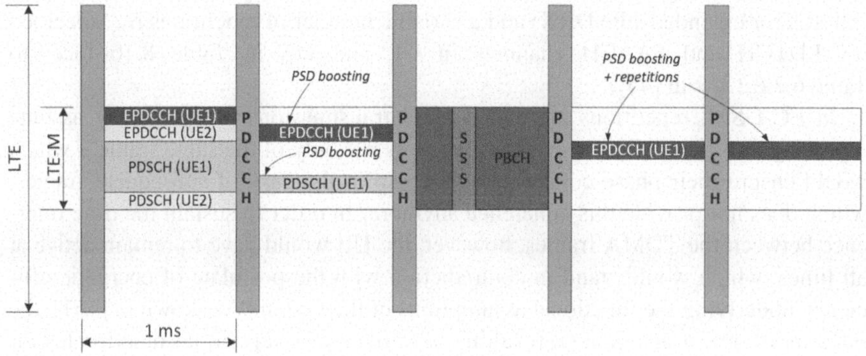

Fig. 8.14 An LTE-M 1.4 MHz channel embedded in the LTE channel [146]

available channel width. Over longer distances, in order to improve the link energy budget, the 'PSD boost' is first used by means of reducing the number of UEs covered by DL transmission and thus increasing the spectral power density for the remaining users. Connectivity at the largest distances, on the other hand, is maintained by implementing, beside the 'PSD boosting', also repetitions.

The UL transmission is shown in Fig. 8.13. Similar to DL, the basic LTE transmission system remains unchanged, with the exception of modifications leading to increased coverage, i.e. 'PSD boosting' and repetitions, for EUs deployed in high-attenuation area locations. Moreover, in UL, as the diagram shows, there is also a possibility of transmission using individual resource elements (REs), i.e. using a channel narrower than the width of an individual physical resource block (PRB) and, in the quoted example, using 4 instead of 12 subcarriers.

A significant feature of LTE-M is the fact that they share the same channel raster with LTE, so that both systems can coexist in the same spectral region without mutual interference. In UL, both systems are frequency-multiplexed (FDM). In DL the Time Division Multiplexing (TDM) is used in the form of a PDCCH channel, as in Fig. 8.14. Naturally, such a multiplexing of two systems transmission in a single subframe requires adequate shortening of the LTE-M downlink channels in the time domain. Moreover, LTE-M requires one empty subcarrier, which leads to a slight shift between physical resource blocks (PRBs) LTE and LTE-M.

8.5 EC-GSM IoT

Extended Coverage (EC)-GSM IoT system is the evolution of 2nd generation standards, i.e. GSM/GPR/EDGE, adapted to the needs of machine-type traffic (MTC), while maintaining all assumptions required from CIoT systems, such as increasing MCL by 20 dB relative to GPRS (where $MCL = 144$ dB), and a significant reduction in the radio module complexity, in order to extend the battery-powered operation time to about 10 years (assuming sporadic transmission). These goals are supposed to be achieved while maintaining compatibility with GSM physical channels and using the following mechanisms: PSM, eDRX (defined only for the idle mode, i.e. e-IDRX, or extended-Idle DRX) and a variable number of repetitions N_{rep} specified for PDTCH and PAACH channels in UL and DL in Table 8.16 (acc. to Table 6.2.6.13–2 in [45]).

In EC-GSM, repetitions are called blind transmissions, because they assume repeating information in subsequent time slots not only of the same frame – which would ensure their phase coherence – but also in the gaps of subsequent frames, which does not provide this coherence anymore. In order to sustain the time reference between the TDMA frames, however, the UE would have to remain active at all times, which would stand in contradiction with the postulate of energetic efficiency underlying the functional assumptions of the system. As shown in [151], the linearity of processing gain G_p resulting from the use of repetitions deteriorates for $N_{rep} > 16$. For example, in an extreme case of $N_{rep} = 1024$, the actual processing gain would therefore equal merely to 24 dB which corresponds to ideally coherent summation obtained already for 256 repetitions.

For the low-energy e-IDRX periods, 12 values for inactivity periods have been proposed, ranging from approx. 1.9 s to approx. 52 min. In terms of modulation, the standard assumes full compliance with the four lowest MCS1-MCS4 schemes defined for the EDGE system, according to Table 8.17. A reference to this system was accepted as binding, due to the fact that in contrast to GPRS, RLC/MAC (Radio Resource Control/Medium Access Control) headers in EDGE are coded using stronger codes than the information content of packets (payload).

In DL, the formats of some control channels have changed, such as PCH (CCCH/D), AGCH (CCCH/D), BCCH and PACCH, modifying them to accommodate different CC coverage classes. These changes affect mostly modifications of indexes and the number of time slots in which these channels are distributed. The

Table 8.16 A number of repetitions depending on the CC cover class

CC# coverage class	N_{rep} in PDTCH i PACCH channels	MCS
CC1	1	MCS-4
CC2	1	MCS-1
CC3	2	MCS-1
CC4	4	MCS-1
CC5	8	MCS-1
CC6	16	MCS-1

Table 8.17 Modulation and code diagrams for EDGE and resulting baud rate

MCS	R_b [kb/s] no RLC/MAC overhead	R_b [kb/s] with RLC/MAC overhead	Modulation	Coding rate
MCS-1	9.20	8.00	GMSK	~0.53
MCS-2	11.60	10.40	GMSK	~0.66
MCS-3	15.20	14.80	GMSK	~0.85
MCS-4	18.00	16.80	GMSK	1

information content carried by these channels has also been reduced, in order to constrain the amount of time necessary for their reception by a UE.

In the UL, only the EC-PACCH/U control channel format and the EC-PDTCH data transmission packet were modified analogously to the changes introduced in DL channels. As an option, a so-called overlaid CDMA (or OLCDMA) [147–149] was added for increasing the UL capacity by multiplexing multiple users simultaneously, using blind retransmissions. It is done by modifying the phase shift in each time slot by $OC{\cdot}\pi$, where the OC multiplier is a row in the binary Hadamard code matrix, whose rows – each assigned to a different user – are mutually orthogonal. As for the physical implementation of OLCDMA, the modulated RF carrier of one user in a given time slot is defined by the formula (8.2), which implies that in each of the N_{rep} slots, the user will repeat his signal shifted by $OC{\cdot}\pi$, according to the binary code assigned to this user, as stated above.

$$x(t') = \sqrt{\frac{2E_b}{T}} \cos\left(2\pi f_0 \left(t' + t_0\right) + \varphi\left(t'\right) + \varphi_{157} + \varphi_0 + OC \cdot \pi\right) \qquad (8.2)$$

where

- t' – transmission time basis (transmission time of the original signal).
- t_0 – time offset equal to the product of the ordinal repetition number and the time slot duration (0.577 μs).
- φ_{157} – phase shift equal to $\pi/2$. If a given retransmission (repetition) is carried out on the zeroth (TS0) or fourth (TS4) slot, then in order to retain the phase continuity, all subsequent retransmissions within the frame will be shifted by φ_{157}, due to the fact that the TS0 and TS4 slots contain 157 symbols [150], which makes them longer than the slots TS1-TS3 and TS5-TS7 containing 156.25 symbols.
- φ_0 – a random phase shift, constant for all blind transmissions within a given TDMA frame.
- E_b – energy per a modulating bit.
- T – bit duration.
- f_0 – centre frequency.
- OC – a π multiplier in the OLCDMA phase shift, as given in Table 8.19 (acc. to [149]).

Due to the use of repetition, on the receiver side, an implemented phase-locked loop (PLL) to effectively decode individual users' signals multiplexed using

Table 8.18 Typical values of UE modem average power consumption

Condition of the device	Description	Power consumption (μA)
Transmission (Tx)	Tx (GMSK) (33 dBm)[1]	1,227,431
	Tx (GMSK) (23 dBm) [2]	152,543
Reception (Rx)	Rx with processing in the baseband	30,000
PLL	Phase synchronization	30,000
Energy-saving modes	Light sleep[3]	1000
	Deep sleep[4]	4.54

[1] Assuming external PA with 50% efficiency (including the Tx/Rx switch insertion losses), increased by 60 mW for other onboard electronics
[2]Assuming external PA with 50% efficiency (including the Tx/Rx switch insertion losses), increased by 60 mW for other onboard electronics
[3] Used in periods between expected transmissions, e.g. between obtaining a synchronization and initiating the random access procedure or receptions, e.g. between the last transmission of the random access procedure and the acquisition of immediate assignment (*IA*). More in Chapter 6.2.6.6.3 of the document [45]
[4] Assuming external PA with 50% efficiency (including the Tx/Rx switch insertion losses), increased by 60 mW for other onboard electronics

OLCDMA, which – as demonstrated in Table 8.18 – affects the increase in UE current consumption. Simulations conducted in [45] and [151] have shown that the maximum number of multiplexed users, achievable without noticeable degradation of transmission quality, is four, for $N_{rep} = 4, 8, 16$. Multiplexation of eight users, with a slight degradation in the reception quality, is possible at $N_{rep} = 16$. For practical applications, however, it is suggested to multiply four users for the EC-PACCH and EC-PDTCH channels, assigning each of them one of the four codes given in Table 8.19. This technique is also used in the EC-RACH (Random Access Channel) in the so-called EDAB mode (*Extended Dual Slot Access Burst*), if the UE supports the coverage class CC5. This mode consists in transmitting RACH queries in two time slots, using the 2-bit code. On the receiving side (in the base station), in order to separate the transmission from a given UE (e.g. using the code #2, i.e. {‘0;0;1;1′}) from transmissions from the three other UEs, all retransmissions received within a given terminal TDMA frames are buffered, and then the signals in individual time slots are conjugate phased-shifted with respect to the user's code. Then, individual

Table 8.19 Codes used in the OLCDMA technique

	OLCDMA code space for the Hadamard matrix values of the *OC* multiplier in the Eq. (8.2)		
OLCDMA code number	Code sequence for channels: EC-PACCH and EC-PDTCH	Code sequence for EC-RACH channel	Code space for the Fourier matrix values of the *OC* multiplier in the Eq. (8.2)
0	{0; 0; 0; 0}	{0; 0}	{0; 1; 0; 1}
1	{0; 1; 0; 1}	{0; 1}	{0; ½; 1; ¾}
2	{0; 0; 1; 1}		{0; 0; 0; 0}
3	{0; 1; 1; 0}		{0; ¾; 1; ½}

I/Q samples are summed up, which leads to coherent accumulation of the desired signal (i.e. intended for the given user) and mutual cancellation of signals from other users. Details of the procedure, presented in the form of an example of signal extraction to a given user, are presented in Appendix 2 (Chap. 11). The case of using blind retransmissions without user multiplexing is equivalent to using the code {0; 0; 0; 0} (for EC-PACCH and EC-PDTCH) and {0; 0} (for EC-RACH). An alternative set of orthogonal codes for use in OLCDMA is a set of lines from the Fourier matrix, presented in Table 8.19, while the multiplexing method itself is analogous to the described procedure for Hadamard codes.

Of the four radiated power levels used in GSM UEs, two levels with the lowest values are used in EC-GSM, namely, the classes: 4th (EIRP = 2 W or 33 dBm) and 5th (EIRP = 0.8 W or 29 dBm). In addition, a class with an EIRP power of 200 mW (23 dBm) was added, which allows the power amplifier (PA) to be integrated in a shared configuration with the radio modem.

The main process at the link layer consists in establishing a connection between a UE and a BS, especially during the Random Access Channel (*RACH*) procedure. Due to the fact that its unsuccessful completion (i.e. the failure to obtain a connection with the BS) would require repetitions, which could be energetically uneconomical to the UE, and would unnecessarily burden the network, the mechanisms used are aimed at obtaining the target interest rate of False Detection Rate (*FDR*) at a level not worse than GSM, i.e. a maximum of 0.02%. As in the case of previously discussed CIoT systems, in EC-GSM there is a division into four coverage classes: [45] (CC1-CC4). Each of them is assigned a number of blind transmissions (see Table 8.20), where CC1 is equivalent to the range of traditional GPRS, while CC4 corresponds to the extreme range for *MCL* = 164 dB. Their usage mechanism assumes that after N_{rep} access requests (*AR*) and failure to obtain the access response (*AR*) from the BS, the UE may switch over to a higher coverage class and during the next access attempt to perform a procedure with a new, increased number of repetitions, corresponding to the new CC.

According to the link energy budget presented in [45] (Chapter 6.2.6.9) for EC-GSM, in terminals with EIRP = 33 dBm, the assumed *MCL* = 164 dB is reachable in logical channels in both directions (UL/DL), EC-PDTCH/U(L) and EC-PACCH/U(L), as well as in DL channels, EC-CCCH/D and EC-BCCH,

Table 8.20 A number of blind transmissions (repetitions) and cover classes for different logical channels [151]

	A number of blind transmissions (coverage class)			
Logical channel	CC1	CC2	CC3	CC4
EC-SCH	28			
EC-BCCH	16			
EC-CCCH/D	1 (CC1)	8 (CC2)	16 (CC3)	32 (CC4)
EC-CCCH/U	1 (CC1)	4 (CC2)	16 (CC3)	**48** (CC4)
EC-PACCH	1 (CC1)	4 (CC2)	8 (CC3)	16 (CC4)
EC-PDTCH	1 (CC1)	4 (CC2)	8 (CC3)	16 (CC4)

assuming the following mobility scenarios: a slow-varying channel with 1 Hz Doppler shift and mobile channel with a 25 Hz shift corresponding to the motion speed of 30 km/h. In the case of terminals with EIRP = 23 dBm, however, the maximum range is limited to the value of MCL = 154 dB. Estimated delays in packet delivery for the UL, in turn, are ~ 0.9 s and ~ 3 s for, respectively, MCL = 144 dB and 154 dB while for DL ~0.9 s and ~3 s, respectively, for MCL = 144 dB and 154 dB.

Wrapping up, functional simplifications used in EC-GSM IoT, translating into a total reduction of 15–20% of the complexity of radio modules relative to traditional GPRS/EDGE modems, include [45]:

- Reduction of EIRP-radiated power classes (defined in [45], Chapter 6.2.1.4.2) to class 4 (2 W, 33 dBm) and class 5 (0.8 W, 29 dBm) and addition of a new power class of 200 mW (23 dBm) that allows integration of a power amplifier in a shared radio module chipset. As a reminder, the GSM/GPRS terminals support also the following power classes: 3rd (5 W, 37 dBm) and 2nd (8 W, 39 dBm). These classes expressed in terms of the power consumption have been demonstrated in Table 8.18.
- Limiting the MCS set to the four lowest ones, i.e. MCS1-MCS4, using a two-state GMSK modulation, as shown in Table 8.16.
- Abandoning the sound transmission functionality, in favour of only data transmission in the Packet Switched (*PS*) mode, which has eliminated the need to support all services related to the Circuit Switched (*CS*) communication domain. This translates into savings in the memory use of approx. 38% for program memory and over 36% for static and dynamic memories. Eliminating the CS domain support has also led to the shrinkage of the modem area by up to 20%.
- Expanding the sleep mechanism, taking into account the following: *a light sleep*, activated between the various steps included between the AR access request, UL transmission, acknowledgement reception and DL message detection and *a deep sleep* activated between the intervals separating individual transmissions in the UL, according to the MAR traffic model (described in Sect. 8.1).
- Smaller number of supported frequency bands, usually reduced to a single band below 1 GHz, which allows to use a single power amplifier: external, for 4th power class (33 dBm), or integrated with a radio module, for the newly added 23 dBm power class.
- Limiting the RLC window size down to 16 blocks instead of 64 GPRS blocks [152].
- Using only 7 out of 60 RLC/MAC messages [152].

Annexes

Annex No. 1

Calculations for the inverse cumulative distribution function $CDF^{-1}(p)$ of a normal probability density, necessary to obtain the coefficients $A(p)$ and $B(p)$ in the formula (18) for BEL attenuation in Sect. 3.5.3, based on the Eq. (1) binding $CDF^{-1}(p)$ to the reverse error function $erf^{-1}(p)$.

$$CDF^{-1}(p) = \sqrt{2} \cdot erf^{-1}(2p-1) \tag{1}$$

where p is the probability, $p \in \langle 0; 1 \rangle$.

K. Staniec, *Radio Interfaces in the Internet of Things Systems*,
https://doi.org/10.1007/978-3-030-44846-2

A MATLAB Code

```
function out = BEL(f1,f2,phi,p,typ)


%f1, f2 - frequencies, respectively initial and final, of the range
expressed in [GHz]
- % phi - the incident wave angle with respect to the building
(value '0' corresponds to the perpendicular incidence)
%p - probability <0; 1>
%type - 't' for traditional buildings (index [0]), 'm' for modern,
thermally efficient buildings (index [1])
%erfinv - the inverse error function erf-1(p)

for ind=1:2
    r=[12.64, 28.19];
    s=[3.72, -3.00];
    t=[0.96, 8.48];
    u=[9.6, 13.5];
    v=[2.0, 3.8];
    w=[9.1, 27.8];
    x=[-3.0, -2.9];
    y=[4.5, 9.4];
    z=[-2.0, -2.1];

    f=f1:0.1:f2;


    F_inv = sqrt(2).*erfinv(2*p-1);
    Lh=r(ind)+s(ind).*log10(f)+t(ind).*(log10(f)).^2;
    Le=0.212.*abs(phi);
    sigma1=u(ind)+v(ind).*log10(f);
    sigma2=y(ind)+z(ind).*log10(f);
    mi1=Lh+Le;
    mi2=w(ind)+x(ind).*log(f);
    C=-3.0;

    A=F_inv.*sigma1+mi1;
    B=F_inv.*sigma2+mi2;

            %LBEL_t - attenuation in traditional buildings
            if(ind==1)  LBEL_t=10*log10(10.^(0.1*A)+10.^(0.1*B)+10.
^(0.1*C));
        end
```

```
                %LBEL_m - attenuation in thermally efficient buildings
(modern)
              if (ind==2) LBEL_m=10*log10(10.^(0.1*A)+10.^(0.1*B)+10.
^(0.1*C));
          end
end

semilogx(f,LBEL_m,'--k','LineWidth',3);
    hold on
semilogx(f,LBEL_t,'k','LineWidth',3);
    ylim([0 60]);
    xlim([0.1 100]);
xlabel('FREQUENCY [GHz]');
ylabel('LOSSES  RESULTING  FROM  PENETRATION  INTO  THE  BUILDING
Ldod [dB]');
grid on;
```

Annex No. 2

The example below describes the process of extracting information intended for a user to whom a code sequence no. 2 from the Hadamard matrix, i.e. {0;0;1;1}, has been allocated. Messages to four users have been multiplexed in the whole transmission, with the use of the OLCDMA technique applied in the EC-GSM IoT system.

Before emitting blind transmissions (repetitions), the signals generated by individual users are shifted in each slot by the value determined by a code assigned to each of them. Each code is a single row in the Hadamard matrix, shown in the column 'In the transmitter' in Table 1. The receiver (the base station), knowing the OLCDMA code assigned to each of the four multiplied users, upon receiving blind transmissions in four consecutive time slots (TS0-TS4) buffers them and then performs in each slot a phase shifting of all received samples by a value conjugate to a given user's original code (see 'Compound shift' column). As a result of this operation, signals (repetitions) transmitted by that user (in this case, a user no. 2) in individual time slots become in-phase, while the other users' signals remain pairwise: in-phase and counter-phase. As a result, summing up I/Q samples causes the other users' signals to be mutually cancelled, since they contain four copies of identical signals in terms of values but pairwisely opposite in phases. The sum of samples of the extracted user's signal, in turn, results in a fourfold accumulation of values, since the signals in each time slot, after the conjugate shift, become all in-phase, which leads to their coherent summation.

The extraction procedure presented here is performed in the receiver four times, separately for each user.

Table 1 Example of a single (2nd) user's signal extraction in the OLCDMA technique

OLCDMA code number	In the transmitter				In the receiver								
					Conjugate shift				Conjugate shift outcome				Result
	TS0	TS1	TS2	TS3	TS0	TS1	TS2	TS3	TS0	TS1	TS2	TS3	
0	0	0	0	0	**0**	**0**	**$-\pi$**	**$-\pi$**	0	0	$-\pi$	$-\pi$	Cancellation
1	0	π	0	π					0	π	$-\pi$	0	Cancellation
2	**0**	**0**	**π**	**π**					**0**	**0**	**0**	**0**	**Accumulation**
3	0	π	π	0					0	π	0	$-\pi$	Cancellation

Bibliography

1. Ashton K (2009) That 'Internet of Things' Thing. In the real world, things matter more than ideas. RFiD J, based on the presentation for Procter & Gamble (P&G) in 1999, June 22, 2009
2. Peter Day's World of Business. BBC World Service (BBC), audio file available on-line at http://downloads.bbc.co.uk/podcasts/radio/worldbiz/worldbiz_20150319-0730a.mp3. Last access date 04.12.2018
3. Huvio E, Grönvall J, Främling K (2002) Tracking and tracing parcels using a distributed computing approach. In: Solem O (ed) Proceedings of the 14th Annual Conference for Nordic Researchers in Logistics (NOFOMA'2002), Trondheim, Norway, 12–14 June 2002, pp 29–43
4. Magrassi P (2002) Why a universal RFID infrastructure would be a good thing. Gartner research report G00106518, 2nd May 2002
5. Magrassi P, Berg T (2002) A world of smart objects. Gartner research report R-17-2243, 12th August 2002
6. Internet of Things—An action plan for Europe. COM (2009) 278 final, Commission of the European Communities, 18th June 2009
7. Wood A (2015) The Internet of Things is revolutionizing our lives, but standards are a must. The Guardian, 31st March 2015
8. Evans D (2011) The Internet of Things: how the next evolution of the internet is changing everything. Cisco White Paper, April 2011
9. IEEE Std 802.15.1-2005, Wireless Medium Access Control (MAC) and Physical Layer (PHY) specification for wireless personal area networks (WPANS), June 2005
10. IEEE, IEEE Std 802.15.4-2006, Part 15.4: Wireless Medium Access Control (MAC) and Physical layer (PHY) Specification for Low-Rate Wireless Personal Area Networks (WPANs)
11. 802.15.6-2012 – IEEE Standard for Local and metropolitan area networks – Part 15.6: Wireless Body Area Networks
12. SmartM2M; IoT Standards landscape and future evolutions. ETSI TR 103 375 V1.1.1 (2016-10)
13. Heiles J (2015) Platforms for connected factories of the future. Alliance for Internet of Things Innovation (AIOTI) workshop. Platforms for connected factories of the future, Brussels, October 5th 2015
14. 3GPP TR 36.785 V14.0.0 (2016-10) 3rd Generation Partnership Project; Technical Specification Group Radio Access Network; Vehicle to Vehicle (V2V) services based on LTE side link; User Equipment (UE) radio transmission and reception (Release 14)
15. 3GPP TR 22.891 V14.2.0 (2016-09) 3rd Generation Partnership Project; Technical Specification Group Services and System Aspects; Feasibility Study on New Services and Markets Technology Enablers; Stage 1 (Release 14)
16. 3GPP TR 22.861 V14.1.0 (2016-09) 3rd Generation Partnership Project; Technical Specification Group Services and System Aspects; Feasibility Study on New Services and Markets Technology Enablers for Massive Internet of Things; Stage 1 (Release 14)

17. ITU-R M.2083 (09/2015), IMT vision – framework and overall objectives of the future development of IMT for 2020 and beyond, November 2014
18. 3GPP TR 22.862 V14.1.0 (2016-09) Technical report 3rd Generation Partnership Project; Technical Specification Group Services and System Aspects; Feasibility Study on New Services and Markets Technology Enablers for Critical Communications; Stage 1 (Release 14)
19. 3GPP TR 22.863 V14.1.0 (2016-09) Technical report 3rd Generation Partnership Project; Technical Specification Group Services and System Aspects; Feasibility Study on New Services and Markets Technology Enablers – Enhanced Mobile wideband; Stage 1 (Release 14)
20. 3GPP TR 22.864 V15.0.0 (2016-09) Technical report 3rd Generation Partnership Project; Technical Specification Group Services and System Aspects; Feasibility Study on New Services and Markets Technology Enablers – Network Operation; Stage 1 (Release 15)
21. Greenough J (2017) 34 billion devices will be connected to the Internet by 2020. Business Insider, February 17, 2017
22. Embracing the Internet of Everything. Cisco, White Paper, 2016
23. A guide to the Internet of Things Infographic. International Data Corporation (IDC), Intel, United Nations, 2015
24. Branderhorst G (2016) IoT and sustainable development – United Nations. Klynveld Peat Marwick Goerdeler, September 2016
25. Internet of Things – number of connected devices worldwide 2015–2025. Statista Research Department, Nov 14, 2019
26. ITU, ITU-T Y.2060 (06/2012) Overview of the Internet of Things. Series Y: global information infrastructure, internet protocol aspects and next-generation networks. Next Generation Networks – frameworks and functional architecture models
27. Buntz B (2016) Top 10 reasons people aren't embracing the IoT. Internet of Things Institute
28. Scully P (2016) 5 things to know about the IoT platform ecosystem. IoT analytics. Market insights for the Internet of Things, January 2016
29. Bartje J (2016) The top 10 IoT application areas – based on real IoT projects The top 10 IoT application areas – based on real IoT projects. IoT analytics. Market insights for the Internet of Things, August 2016
30. Zelkha E, Epstein B, Birrell S, Dodsworth C (1998) From devices to "Ambient intelligence", Digital living room conference, June 1998
31. McCullough M (2005) Digital ground: architecture, pervasive computing, and environmental knowing. MIT Press
32. Surie D (2012) Egocentric interaction for ambient intelligence. PhD. thesis, Department of Computing Science, Umeå University, Sweden
33. Guo B, Zhang D, Imai M Towards a cooperative programming framework for context-aware applications. ACM/Springer J Pers Ubiquit Comput 15(3):221–233
34. Bieliková M, Krajcovic T (2001) Ambient intelligence within a home environment. ERCIM News, October 2001
35. Nikhil P. Ambient intelligence. IBM, 2013. The document available at the address: https://www.slideshare.net/nikhilpatteri/ambient-intelligence-28697238. Last access date: 29.11.2017
36. Fettweis G et al (2014) The tactile internet. ITU-T technology watch report, August 2014
37. IEEE P1918.1 (2016) Tactile internet: application scenarios, definitions and terminology, architecture, functions, and technical assumptions
38. Deshmukh UA, More SA (2016) Fog computing: new approach in the world of cloud computing. Int J Innov Res Comp Commun Eng 4(9):16310–16316
39. Satyanarayanan M (2017) The emergence of edge computing. IEEE Computer Society:30–39
40. MacGillivray C, et al (2016) IDC analyze the future, IDC FutureScape: Worldwide Internet of Things 2017 predictions. IDC #US40755816e, November 2016
41. Fog computing and the Internet of Things: extend the cloud to where the things are. Cisco Whitepaper C11-734435-00, 2015

42. Cloud, fog and edge computing – what's the difference? WinSystems. The Embedded Systems Authority, 2017. The document available at the address https://www.winsystems.com/cloud-fog-and-edge-computing-whats-the-difference/. Last access date: 28.12.2017
43. IQRF. Technology for wireless. The document available at the address: https://www.iqrf.org/. Last access date: 05.12.2017
44. 3GPP TR 36.888. Study on provision of low-cost Machine-Type Communications (MTC) User Equipments (UEs) based on LTE, v.12.0.0, June 2013
45. TR 45.820. Cellular system support for ultra low complexity and low throughput Internet of Things. V2.1.0, August, 2015
46. Directive 2006/32/EC of the European Parliament and of the Council of 5 April 2006 on energy end-use efficiency and energy services and repealing Council Directive 93/76/EEC
47. European Commission Standardisation Mandate M/441: "Standardisation mandate to CEN, CENELEC and ETSI in the field of measuring instruments for the development of an open architecture for utility meters involving communication protocols enabling interoperability", Brussels, 12th March 2009
48. Machine-to-Machine communications (M2M); M2M service requirements, ETSI TS 102 689 V2.1.1 (2013-07)
49. Machine-to-Machine communications (M2M); Functional architecture, ETSI TS 102 690 V1.2.1 (2013-06)
50. Antón-Haro C, Dohler M (2015) Machine-to-machine (M2M) communications. Architecture, performance and applications. Woodhead Publishing Series in Electronic and Optical Materials: number 69, Elsevier
51. Machine-to-Machine communications (M2M); Use cases of automotive applications in M2M capable networks, ETSI TS 102 898 V1.1.1 (2013-04)
52. Machine-to-Machine communications (M2M); Smart Metering Use Cases, ETSI TR 102 691 V1.1.1 (2010-05)
53. Machine-to-Machine communications (M2M); mIa, dIa and mId interfaces, ETSI TS 102 921 V1.1.1 (2012-02)
54. IEEE P2413 – Standard for an architectural framework for the Internet of Things (IoT)
55. International Standard ISO/IEC/IEEE42010. Systems and software engineering – architecture description, 1st edn, 2011-12-01
56. CEN, EN 13757-1:2002. Communication system for meters and remote reading of meters – part 1: data exchange. European Committee for Standardization, 07.11.2002
57. CEN, EN 13757-2:2004. Communication systems for meters and remote reading of meters – part 2: physical and link layer. European Committee for Standardization, 24.11.2004
58. CEN, EN 13757-3:2013. Communication systems for meters and remote reading of meters – part 3: dedicated application layer. European Committee for Standardization, 07.03.2013
59. CEN, EN 13757-4:2013. Communication systems for meters and remote reading of meters – part 4: wireless meter readout (Radio meter reading for operation in SRD bands). European Committee for Standardization, 29.06.2013
60. Antipolis S (2014) Billions of connected objects now have the solution for low throughput connectivity. European Telecommunications Standards Institute (www.etsi.org)
61. Low Throughput Networks (LTN); Use cases for Low Throughput Networks, ETSI GS LTN 001 V1.1.1 (2014-09)
62. Myers T (2010) Random phase multiple access communication interface system and method. U.S. Patent 7 782 926 B2, Aug. 24, 2010
63. Low Throughput Networks (LTN); Functional architecture, ETSI GS LTN 002 V1.1.1 (2014-09)
64. Hata M (1980) Empirical formula for propagation loss in land mobile radio services. IEEE Trans Veh Technol 29(3):317–325
65. Cichon DJ, Kürner T. Propagation prediction models", ch. 4.4: „Propagation models for macrocells", COST 231 final report D.J. Cichon, T. Kürner, "Propagation prediction models", ch. 4.7: „Indoor propagation models", COST 231 final report

66. Meinilä J et al (2010) Wireless World Initiative New Radio – WINNER+. D5.3: WINNER+ Final channel models. CELTIC/CP5-026, Heino P (ed)
67. Kyösti P. WINNER II channel models. IST-4-027756 WINNER II D1.1.2 V1.2, 2008
68. Rudd R et al (2014) Building materials and propagation. Final report. Ofcom, 2604/BMEM/R/3/2.0
69. ITU, ITU-R P.2108-0 (06/2017). Prediction of clutter loss
70. ITU, ITU-R P.526-8. Propagation by diffraction
71. Mobile broadcast technologies. Link budgets. Broadcast mobile convergence BMCOForum, 02/2009
72. Okumura Y et al (1968) Field strength and its variability in the VHF and UHF land mobile radio service. Rev Elec Community Lab 16(9/10):825–893
73. ITU, ITU-R P.2346-2 (03/2017) Compilation of measurement data relating to building entry loss
74. ITU, ITU-R P.2040-1 (07/2015) Effects of building materials and structures on radiowave propagation above about 100 MHz
75. ITU, ITU-R P.2109-0 (06/2017) Prediction of building entry loss
76. ANSI/IEEE Std 802.11, 1999 Edition, Information technology – Telecommunications and information exchange between systems – Local and metropolitan area networks – Specific requirements – part 11: Wireless LAN Medium Access Control (MAC) and Physical Layer (PHY) Specifications
77. ETSI TR 103 435 V1.1.1 (2017-02) System Reference document (SRdoc); Short Range Devices (SRD); Technical characteristics for Ultra Narrow Band (UNB) SRDs operating in the UHF spectrum below 1 GHz
78. Low Throughput Networks (LTN); Protocols and Interfaces, ETSI GS LTN 003 V1.1.1 (2014-09)
79. ETSI TR 102 313 V1.1.1 (2004-07) Electromagnetic compatibility and Radio spectrum Matters (ERM); Frequency-agile Generic Short Range Devices using Listen-Before-Transmit (LBT); Technical report
80. ETSI EN 300 220-1 V3.1.1 (2017-02) Short Range Devices (SRD) operating in the frequency range 25 MHz to 1 000 MHz; Part 1: technical characteristics and methods of measurement
81. ETSI EN 300 220-2 V3.1.1 (2016-11) Short Range Devices (SRD) operating in the frequency range 25 MHz to 1 000 MHz; Part 2: Harmonised Standard covering the essential requirements of article 3.2 of Directive 2014/53/EU for non specific radio equipment
82. The European table of frequency allocations and applications in the frequency range 8.3 kHz to 3000 GHz (ECA Table), approved in October 2017
83. Directive 2014/53 /EU of the European Parliament and of the Council of 16 April 2014 on the harmonization of the laws of the Member States concerning the making available on the market of radio equipment and repealing Directive 1999/5/EC
84. Winkler M (1962) Chirp signals for communications. IEEE WESCON Conv Rec Pt. 7
85. Flandrin P (1999) Time-frequency/time-scale analysis. Academic, San Diego. ISBN 0-12-259670-9
86. Springer A et al (2000) Spread spectrum communications using chirp signals. IEEE Proc Eurocomm:166–170
87. Goursaud Z, Gorce J-M (2015) Dedicated networks for IoT: PHY/MAC state of the art and challenges. EAI endorsed transactions on Internet of Things, HAL Id: hal-01231221
88. IEEE 802.15.4-2011: "IEEE Standard for Local and metropolitan area networks – part 15.4: Low-Rate Wireless Personal Area Networks (LR-WPANs)"
89. Paret D, Huon J-P (2017) Secure connected objects. ISTE Ltd/Wiley, London
90. Cook CE, Bernfeld M (1967) Radar signals – an introduction to theory and applications. Academic, New York
91. Do M-T, Goursaud Z, Gorce J-M Interference modelling and analysis of random FDMA scheme in ultra narrowband networks. In: International conference on application of information and communication technologies (AICT), July 2014, Paris, France

92. Win MZ, Pinto PC, Shepp LA (Feb. 2009) A mathematical theory of networks interference and its applications. Proc IEEE 97(2):205–230
93. Saunders SR, Aragón-Zavala A (2007) Antennas and propagation for wireless communications systems, 2nd edn. Wiley, New York
94. ITU, ITU-R P.1407-1. Multipath propagation and parameterization of its characteristics
95. Parsons JD (2000) Mobile radio propagation channel, 2nd edn. Wiley, New York
96. Do M-T, Goursaud Z, Gorce J-M (2014) On the benefits of random FDMA schemes in ultra narrow band networks. In: IEEE 12th International Symposium on Modeling and Optimization in Mobile Ad Hoc, and Wireless Networks (WiOpt), pp 672–667
97. ERC Recommendation 70-03. Relating to the use of Short Range Devices (SRD). Tromsø, 19 May 2017
98. 3GPP TR 38.900 V15.0.0 (2018-06) 3rd Generation Partnership Project; Technical Specification Group Radio Access Network; Study on channel model for frequency spectrum above 6 GHz (Release 15)
99. LoRa modulation basics. Semtech, Application Note AN1200.22
100. Sornin N, Luis M, Eirich T, Kramp T, Hersent O (2016) LoRaWAN Specification, Version:V1.0.2, July 2016
101. Aloÿs A et al (2016) A study of LoRa: long range & low power networks for the Internet of Things. Sensors 16:1–18
102. LoRa Alliance Technical committee. LoRaWAN Regional Parameters, Version: V1.0, July 2016
103. Nanoscale Labs. Communications system. Patent US8406275, submitted 1020-03-09, published 2012-03-26
104. Ray B LPWA in an LTE-M and NB-IoT World. Link Labs LPWA Webinar, Nov. 9th, 2016
105. SX 1272/3/6/7/8: LoRa modem. Semtech, Designer's guide, AN1200.13, rev 1, July 2013
106. Weightless(-P). System specification, version 1.03, 7th November 2017
107. https://www.sigfox.com/en
108. SigFox. Technical overview. SigFox, May 2017. Electronic document available at the address: https://www.disk91.com/wp-content/uploads/2017/05/4967675830228422064.pdf. Last access on 06.12.2018
109. Palaios A, Miteva V, Riihijärvi J, Mähönen P (2016) When the whispers become noise: a contemporary look at radio noise levels. 2016 IEEE Wireless Communications and Networking Conference (WCNC), 3–6 April 2016. https://doi.org/10.1109/WCNC.2016.7564891
110. Chang M-H, Lin K-H (2004) A comparative investigation on urban radio noise at several specific measured areas and its applications for communications. IEEE Trans Broadcast 50(3):233–243
111. Vejlgaard B, Lauridsen M, Nguyen HC, Kovács I, Mogensen PE, Sørensen M (2017) Interference impact on coverage and capacity for low power wide area IoT networks. In: IEEE Wireless Communications and Networking Conference 2017
112. 3GPP TR 22.861 V14.1.0 (2016-09) 3rd Generation Partnership Project; Technical Specification Group Services and System Aspects; Feasibility Study on New Services and Markets Technology Enablers for Massive Internet of Things; Stage 1 (Release 14)
113. ITU, ITU-R P. 1407-6 (06/2017) Multipath propagation and parameterization of its characteristics. P Series: radiowave propagation
114. Research group of Prof. Paul Walter Baier, U. of Kaiserslautern, Germany
115. Pomianek AJ, Staniec K, Jóskiewicz Z (2010) Practical remarks on measurement and simulation methods to emulate the wireless channel in the reverberation chamber. Prog Electromagn Res (PIER) 105:49–69
116. Hill DA (2009) Electromagnetic fields in cavities: deterministic and statistical theories, The IEEE Press series on electromagnetic wave theory. Wiley
117. IEC 61000-4-21 Electromagnetic compatibility (EMC) Part 4: Testing and measurement techniques Reverberation chamber test methods. Section 21: reverberation chamber test methods

118. Staniec K, Kowal M LoRa performance under variable interference and heavy-multipath conditions. Wirel Commun Mob Comput 2018:1–9. https://doi.org/10.1155/2018/6931083
119. Ratasuk R, Mangalvedhe N, Ghosh A (2015, September) Overview of LTE enhancements for cellular IoT. In: 26th international symposium on personal, indoor and mobile radio communications, Hong Kong, China, pp 2293–2297
120. Ratasuk R, Prasad A, Li Z, Ghosh A, Uusitalo M (2015, February) Recent advancements in M2M communications in 4G networks and evolution towards 5G. In: 18th conference on intelligence in next generation networks: innovations in services, networks and clouds (ICIN'15)
121. Rapeepat R, Vejlgaard B, Nitin M, Amitava G NB-IoT system for M2M communication. In: 2016 IEEE Wireless Communications and Networking Conference Workshops (WCNCW), Doha, Qatar, 3–6 April 2016
122. Eric WY-P et al (2017) A primer on 3GPP narrowband Internet of Things. IEEE Commun Mag 55(3):117–123
123. Shariatmadari H et al (2015) Machine-type communications: current status and future perspectives toward 5G systems. IEEE Commun Mag 53(9):10–17
124. 3GPP TR 38.913 V14.3.0. Study on scenarios and requirements for next generation access technologies
125. ITU-R M.2412-0. Guidelines for evaluation of radio interface technologies for IMT-2020
126. ITU-R M.2410-0. Minimum requirements related to technical performance for IMT-2020 radio interface(s)
127. R1-157248. NB IoT – Capacity evaluation. Nokia networks, RAN1#83, Anaheim, USA, 2015
128. Coverage analysis of LTE-M category-M1, document developed by: Sierra Wireless, Ericsson, Nokia, Sony, Docomo NTT i in., Version 1.0, January 2017 (available on-line)
129. USCH simulation summary for Rel-13 LC UEs, R1-150759, Sierra Wireless
130. Cross PRB Channel Estimation for M-PDCCH, R1–154211, Sony
131. PUSCH channel estimation aspects for MTC, R1-151216, Ericsson
132. 3GPP TS 36.101. User Equipment (UE) Radio Transmission and Reception. 3rd Generation Partnership Project; Technical Specification Group Radio Access Network; Evolved Universal Terrestrial Radio Access (E-UTRA)
133. 3GPP TS 36.104. Base Station (BS) radio transmission and reception. 3rd Generation Partnership Project; Technical Specification Group Radio Access Network; Evolved Universal Terrestrial Radio Access (E-UTRA)
134. 3GPP TS 23.682 V16.1.0 (2018-12) Technical Specification Group Services and System Aspects; Architecture enhancements to facilitate Communications with packet data networks and applications (Release 16)
135. 3GPP TS 24.301 V15.5.0 (2018-12) Technical Specification Group Core Network and Terminals; Non-Access-Stratum (NAS) protocol for Evolved Packet System (EPS); Stage 3 (Release 15)
136. 3GPP TS 23.401 V16.1.0 (2018-12) 3rd Generation Partnership Project; Technical Specification Group Services and System Aspects; General Packet Radio Service (GPRS) enhancements for Evolved Universal Terrestrial Radio Access Network (E-UTRAN) access (Release 16)
137. NB-IoT deployment guide to basic feature set requirements. CLP.28, GSM Association, Version 1.0, 2nd August 2017
138. 3GPP TS 36.304 V15.2.0 (2018-12) 3rd Generation Partnership Project; Technical Specification Group Radio Access Network; Evolved Universal Terrestrial Radio Access (E-UTRA); User Equipment (UE) procedures in idle mode (Release 15)
139. 3GPP TS 36.321 V15.4.0 (2018-12) 3rd Generation Partnership Project; Technical Specification Group Radio Access Network; Evolved Universal Terrestrial Radio Access (E-UTRA); Medium Access Control (MAC) protocol specification (Release 15)
140. 3GPP TS 24.008 V15.5.0 (2018-12) Technical Specification Group Core Network and Terminals; Mobile radio interface Layer 3 specification; Core network protocols; Stage 3 (Release 15)

141. ETSI TS 136 211 V14.2.0 (2017-04) LTE; Evolved Universal Terrestrial Radio Access (E-UTRA); Physical channels and modulation (3GPP TS 36.211 version 14.2.0 Release 14)
142. 3GPP TS 36.213, Evolved Universal Terrestrial Radio Access (E-UTRA); Physical layer procedures. Release 15 (modified in May 2018 r)
143. ETSI TR 136 942 V15.0.0 (2018-07) LTE; Evolved Universal Terrestrial Radio Access (E-UTRA); Radio Frequency (RF) system scenarios (3GPP TR 36.942 version 15.0.0 Release 15)
144. ETSI TS 144 065 V15.0.0 (2018-07) Digital cellular telecommunications system (Phase 2+) (GSM); Mobile Station (MS) – Serving GPRS Support Node (SGSN); Subnetwork Dependent Convergence Protocol (SNDCP) (3GPP TS 44.065 version 15.0.0 Release 15
145. Evaluation of LTE-M towards 5G IoT requirements, document developed by: Sierra Wireless, Ericsson, Nokia, Sony, Docomo NTT i in., Version 1.1, March 2018 (available on-line)
146. Nokia – LTE M2M: optimizing LTE for the Internet of Things. Nokia White Paper, Global mobile Suppliers Association, 2015 (available on-line)
147. ETSI TS 145 002 V15.1.0 (2018-10) Digital cellular telecommunications system (Phase 2+) (GSM); GSM/EDGE Multiplexing and multiple access on the radio path (3GPP TS 45.002 version 15.1.0 Release 15)
148. ETSI TS 145 004 V13.1.0 (2016-04) Digital cellular telecommunications system (Phase 2+) (GSM); Modulation" (3GPP TS 45.004 version 13.1.0 Release 13)
149. 3GPP TSG GERAN#69, EC-EGPRS, Overlaid CDMA design and performance evaluation, Malta 15–18 Feb., 2016, Ericsson LM, GP-160096
150. ETSI TS 145 010 V10.1.0 (2011-04) Digital cellular telecommunications system (Phase 2+); Radio subsystem synchronization (3GPP TS 45.010 version 10.1.0 Release 10)
151. Liberg O et al (2017) Cellular Internet of Things. Technologies, standards and performance, 1st edn. Elsevier, Academic
152. 3GPP TS 44.060 V15.0.0 (2018-06) 3rd Generation Partnership Project; Technical Specification Group Radio Access Network; General Packet Radio Service (GPRS); Mobile Station (MS) – Base Station System (BSS) interface; Radio Link Control/Medium Access Control (RLC/MAC) protocol (Release 15)

Index

A
Access requests (AR), 177
Activation by personalization (ABP), 100
Adaptive Frequency Agility (AFA), 59
Additive white Gaussian noise (AWGN), 119, 123, 126, 127, 130
Ambient intelligent (AmI)
  challenges, 20
  collaboration, 19
  concepts, 19
  conventional applications, 22
  definition, 19
  distributed computing, 20
  HCI interface
    context awareness, 21
    natural, 22
  home environment, 19
  information, 21
  monitoring, 19
  reasoning functionality
    activity prediction and recognition, 21
    decision-making, 21
    space-time inferring, 21
    user profiling, 21
  user's environment, 22
Anechoic chamber, 121, 122, 128
Attenuation, 61, 66, 70, 72–75
Augmented reality (AR), 22
Automatic repeat request (ARQ), 58

B
Backhaul networks, 36
Base station (BS), 42, 92, 142
Beacon_interval, 96
Blind transmissions, 174
Boltzmann constant, 79
Building entry loss (BEL), 75

C
Capillary networks
  basic, 29
  data transmission, 28–29
  definition, 28
  extended, 29
  media gateways, 29
  traditional radio systems, 28
Carrier sense multiple access/collision detection (CSMA/CA), 58
Carrier to noise and interference (CNIR), 120
Carrier-to-noise ratio (CNR), 123
Cellular IoT (CIoT) systems, 46
Central Registration Authority (CRA), 55
Chirp spread spectrum (CSS), 100
  autocorrelation, 84
  chirp-type signals, 82
  exponential type, 82
  frequency, 82, 83
  LPWAN systems, 82
  parameters, 82
  properties, 84
  quadratic function, 83
  synchronization, 85
  transmission efficiency, 85
  transmission spectral width, 83
CIoT systems
  configurations, 153
  delay, 153
  eDRX, 157–160

K. Staniec, *Radio Interfaces in the Internet of Things Systems*,
https://doi.org/10.1007/978-3-030-44846-2

CIoT systems (*cont.*)
  energy efficiency, 155
  EPA and ETU, 155
  5G mobile systems, 153
  IoT-enabled devices, 153
  LTE, 155
  MAR model, 153
  mMTC scenario, 153
  MTC traffic, 154
  PSM, 156–157
  SNDCP protocol, 153
Circuit Switched (CS), 178
Clear channel assessment (CCA), 56, 58
Cloud computing
  characteristics
    EaaS, 23
    IaaS, 23
    PaaS, 23
    PAYG, 23
    SaaS, 23
  definition, 23
  differentiated options, 23
  Internet availability, 24
  traditional architecture, 25
  types
    hybrid, 24
    private, 23
    public, 24
Colloquially, 130
Command identifier (CID), 99
Connected DRX (C-eDRX), 157, 158
Connected Mode (CM), 164
Control channels, 174
Coverage classes (CC), 161, 166, 169
Customer service platform (CSP), 42
Cyclic redundancy check (PHDR_CRC), 97

D
Delay spread, 130, 131
Dense MTC traffic, 17
Device under test (DUT)
  antennas, 121
  channel, 126
  devices, 132
  low-loss coaxial cables, 121
  OM, 127
  receiver, 133, 134
  signal generator, 126
  transmitter, 126, 133, 134
Direct sequence spread spectrum (DSSS)
  methods, 81
  power spectral density, 80
  XOR, 80
DL traffic transmission, 53
Downlink (DL), 112, 170
Downlink (DL) radio interface, 90
Duty cycle (DC), 56, 110
Dynamic range, 146, 148, 149

E
EC-GSM IoT system
  binary code, 175
  blind retransmissions, 177
  blind transmissions, 174, 177
  control channels, 174, 175
  EC-GSM, classes, 177
  energy budget, 177
  functional assumptions, 174
  GPRS/EDGE modems, 178
  Hadamard code matrix, 175
  low-energy e-IDRX, 174
  PLL, 175
  RACH procedure, 177
  second-generation standards, 174
  TDMA frames, 176
  UL and DL, 178
Edge intelligence
  decisions, 24
  definition, 24
  fog computing, 26
  guidelines
    data reliability, 25
    data safety, 25
    delay minimization, 25
    saving network capacity, 25
Education as a Service (EaaS), 23
Effective isotropic radiated power (EIRP), 94, 99, 100, 108, 113
Effective radiated power (ERP), 56
Electromagnetic Compatibility Laboratory (ECL), 138
Electromagnetic Compatibility Laboratory of the Wrocław University of Science and Technology (ECL, WUST), 121
Electromagnetic interference
  automation systems, 120
  Bluetooth, 120
  groups, 119
  immunity measurement cycle, 121 (*see* EM interference immunity measurement)
  IoT systems, 119
  ISM band, 120
  LAP stations, 119
  SigFox message, 120
EM interference immunity measurement
  CNR requirements
    interference system, 126

receiver, 124
transmitter, 124
extremely multipath propagation, 129
hardware setting tuning
aim, 123
ARF, 124
CNIR curves, 123
OM, 123
SF, 124
PER
interfering setup, 127
receiver, 127, 128
preparation
antennas arrangement, 121
inferencing system, 123
interfering antenna location, 123
receiver system, 123
transmitter module, 123
WUST ECL infrastructure, 121
End devices (ED)
activation, 100
class A devices, 95, 96
class B devices, 96
class C devices, 96, 97
Energy saving, 39
Enhanced MTC (eMTC), 152
Environment
ITU, 72
pathloss, 66
signal sources, 73–74
transmission and reception, 72
urban, 65, 69
variants, 71
ETSI TR 103435 specification, 47
eXclusive OR (XOR), 80
Extended 3GPP model
5G cellular systems, 63
LOS conditions probability, 70
pathloss, 63, 66–70
Extended discontinuous reception mode (eDRX)
C-eDRX, 158
DRX mixed mode algorithm, 160
idle state, 157
I-eDRX, 157, 158
long DRX cycle, 159
operational principle, 159
PDCCH channel, 159
Extended Pedestrian A model (EPA), 155
Extended Typical Urban (ETU), 155
Extremely multipath propagation
AWGN radio channel, 130
colloquially, 130
components, 129
LOS and NLOS, 129
measurement preparation
PER measurement, 134, 135
transceiver system antennas, 132–134
measurement tuning
transmitter, receiver, 134
multiple interactions, 129
PDP, 129–132
urban macrocells, 131

F
False Detection Rate (FDR), 177
Fast Fourier transform (FFT), 101
Fog computing, 2
Fog nodes
Big Data analyses, 26
calculation layers, 27
cloud calculations, 27
cloud layer, 28
data analysis location, 26
definition, 26
real-time control and analysis, 27
smart grid, 26
Forward error correction (FEC), 58
Frequency allocation, 56
Frequency bands, 120
Frequency division-multiplexed (FDM), 173
Frequency-hopping spread spectrum (FHSS), 81
FSK, 93, 106, 108, 112

G
GMSK, 143, 145, 146
GPRS/EDGE modems, 178
Guard band' scenario, 162

H
Hadamard code matrix, 175
Hamming distance, 103
Hata model, 61, 62
High-energy link budget, 39
Hybrid ARQ (HARQ), 162
Hyper Frame Number (HFN), 158

I
Idle DRX (I-eDRX), 157, 158
IEC 61000–4-21, 134
IEEE 802.11 specification, 86
IEEE 802.15.4 standard, 2
IEEE P2413 standard
components, 37
domain and abstraction, 37

IEEE P2413 standard (*cont.*)
IoT architecture, 36
ISO/IEC/IEEE 42010 standard, 37
multi-tier IoT systems, 37
purpose, 37
telecommunication technologies, 37
IMT-2020 systems, 22
In-band' scenario, 161
Indoors
loss standard deviation, 66
probagated, 66
users, 71
Information and Communication Technologies (ICT), 7
Information element (IE), 158
Infrastructure as a Service (IaaS), 23
Interference-free transmission, 88
Internet of Things (IoT)
AmI (*see* Ambient intelligent (AmI))
applications, 8
intelligent agriculture, 3
intelligent environment, 3
intelligent industry, 3
intelligent transport/automotive, 3
intelligent WBAN devices, 3
smart cities, 3
Bluetooth, 2
capillary (*see* Capillary networks)
characteristic
dynamic changes, 10
heterogenity, 10
interconnectivity, 10
large scale, 10
thing-related services, 10
coarse division, 2
development, 17
device, 8, 9
5G, 5
fog computing, 2
history, 1
ICT, 7, 16
intelligent industry domain, 18
intelligent media measurements, 4
ITU Y.2060, 6
M2M, 1, 7
mass-type/critical traffic, 5
reference model (*see* IoT systems reference model)
requirements
autonomous network organization, 10
autonomous service provision, 10
high-quality services, 11
identification-based interconnectivity, 10
location-based services, 11
mangeability, 11
plug and play, 11
privacy protection, 11
security, 11
risk factors (*see* IoT systems risk factors)
technical description, 7
telecommunication traffic, 17
traffic models, 32
transmission parameters, 5
vertical and horizontal domains, 2
Internet Protocol (IP), 24
Inter-site Distance (ISD), 32
Inter-symbol interference (ISI), 130
IoT Architecture, 36
IoT networks standardization
IEEE P2413 standard, 36–38
LTN (*see* Low-Throughput Networks (LTN))
medium access mechanism, 58–60
smart metering, 33–36
spectral issues, 56
IoT systems reference model
applications, 11
hardware layer
device functions, 12
gate functions, 13
management functions, 13
security functions, 14
network layer, 12
service and application support layer, 11
IoT systems risk factors
data security, 14
immature technology, 16
incompatibility, 16
lack of interoperability, 15
less standards, 15
proper infrastructure, 15
restricted knowledge, 14
significant implementation costs, 14
transmission security, 14
uncertainty, 15
ISO/IEC/IEEE 42010 standard, 37

L

Line-of-sight (LOS), 129
Listen Before Talk' (LBT) technique, 59, 60
Long DRX cycle, 158, 160
Long-range LPWAN/LTN/CIoT networks, 30
Long-range wide area network (LoRaWAN/Lora)
characterization, 93
commands, 97

device classes, 95–97
framework structure, 97–100
multi-valued CSS method, 93
physical layer (*see* LoRa physical layer)
preamble, 94
system architecture, 94, 95
LoRa orthogonality, 101
LoRa physical layer
aggregated transmission capacity, 106
CSS, 100
IoT applications, 100
logical channels, 106
MCL signal dynamics, 108, 110
modulation and demodulation, 101–103
packet transmission length, 107
parameters, 106
PER, 106
processing gain, 107, 108
sensitivity, 108
SNR, 107, 108
transmission rate, 104, 105
LoRa system
immunity, multipath propagation
area of susceptibility, 141
BW channel, 141
BW impact, 139
planning principle, 140
spread factor SF, 139
spreading factors, 139
unevenness, 140
immunity, noise and interference
conclusion I, 138
conclusion II, 139
conclusion III, 139
ECL, 138
SNR ratio, 137
sources, 137
wideband interference, 137
LOS conditions probability, 70
Low-Throughput Networks (LTN), 2
application scenarios (*see* LTN applications)
challenges, 41
energy saving, 39
ETSI standardization, 40
functional architecture (*see* LTN functional architecture)
fundamental feature, 41
harsh propagation environments, 40
high-energy link budget, 39
low-bandwidth transmissions, 39
M2M connections, 40
real radio coverage, 38
servicing and maintenance costs, 39
wireless M-Bus, 40
LPWAN network, 85
LTE cellular networks, 151
LTE-M(TC)
DL transmission, 172
FDM, 173
MCL, 172
NB-IoT, 152, 153
PSD boosting, 172, 173
shared channels, 172
subsystem, 151
traditional, 170, 171
LTN Access Point (LAP), 52
LTN applications
automative industry, 48
back-end system, 42
CIoT systems, 46
data transmission, 41
ETSI TR 103435 specification, 41, 47
frame data field sizes, 43
high delay-tolerant M2M traffic, 41
intelligent media measurements, 49, 50
LEP, 42
parameters, 43–45
smart cities, 47
smart grid, 48, 49
transmission parameters, 46
LTN end points (LEP), 42
LTN functional architecture
DL traffic transmission, 53
generic network structure, 50
implementation techniques, 53
interfaces, 52, 53
LAP, 52
message management module, 52
NM, 52
random modem, 52
LEP, 53
LEP terminals, 50, 51
LTN server
message processing, 52
NM, 52
network configurations, 55
radio network, 51
two-directional communication, 53
LTN networks, 79

M

M2M communication applications, 151
M2M gateways
cooperation and integration, 29
data aggregation, 29

M2M gateways (*cont.*)
end device management, 30
local data archiving, 30
reliable communication, 29
securing data and networks, 29
M2M structure
access networks, 35
applications, 36
Backhaul networks, 36
device, 34
gate, 34
local networks, 35
service layers, 36
Machine-to-machine (M2M), 1, 7, 29
Machine-type communication (MTC), 1, 158
Machine-type devices (MTDs), 1
Machine-type traffic (MTC), 149, 151, 174
MACPayload, 98
Macrocell, 63, 66
Main Information Block (MIB), 158
Massive machine-type communication (mMTC), 2
Maximum coupling loss (MCL), 79
Microcell, 63, 66
Middle Eastern and African (MEA) countries, 18
Mobile autonomous reporting (MAR) model, 32, 153, 155
Mobile Station (MS), 42
Modem complexity, 151
Modulation and demodulation
FFT, 101
LoRa chirp signal, 101
LoRa principle, 103
m-state, 101
multistate, 101
orthogonality, 101
preamble and synchronization, 101
process, 103
quadratic function, 101
signal upchirp, 101
Modulation-code schemes (MCS), 168
Multiaccess control (MAC), 86
Multi-access procedure, 153
Multiplexation, 176

N
Narrowband (NB), 112, 144
Narrowband GSM (N-GSM), 153
Narrowband IoT (NB-IoT)
AT, 164
battery lifetime, 166
CC, 169, 170
classes, 168, 169
DL and UL, 170
DL limks, 163
EIRP, 167
eNB, 164
energy consumption, 163
extended indoor coverage, 160, 161
HARQ confirmations, 165
implementation scenarios, 161, 162
ISO/OSI model, 170
long-term operation, 166
MCS, 168
NPSCH channels, 166
OFDM symbols, 165
PDCCH configuration, 166
PDCCH messages, 164
physical channels, 164
power consumption, 163
PRB, 166
PSD boosting, 167
Release 13 (Rel13), 160
scalability, 168
stations, 162
transceiver complexity, 163
transport and physical channels, 165
UE, modes, 164
Narrowband M2M (NB M2M), 153
narrowband-OFDMA (NB-OFDMA), 153
Natural interface, 22
Network community, 1
Network management (NM), 52
Node B (NB), 42
Non-line-of-sight (NLOS), 129
NPSCH channel, 166

O
OpenFog project, 15
Operating Mode (OM), 123, 124, 126, 135
Operating system, 38
Operations Support Systems (OSS), 52
OQPSK, 143, 145, 146
Orthogonal Sequence Spread Spectrum (OSSS), 51
OSSS radio interface, 86
Overlaid CDMA (OLCMDA), 175
Over-the-air radio activation (OTAA), 100

P
Packet Error Rate (PER), 106, 137
Packet Switched (PS) mode, 178
Pathloss, 63, 66, 70
Peak-to-average power ratio (PAPR), 167
Phase-locked loop (PLL), 175
PHYPayloadCRC data field, 98

Physical Downlink Control Channel (PDCCH), 169
Platform as a Service (PaaS), 23
Polite medium access (PMA), 56
Power amplifiers (PA), 167
Power Delay Profile (PDP), 129, 131, 132
Power saving mode (PSM)
  energy saving, 157
  operational procedure, 156
  UE, 156
Preamble
  SyncWord, 101, 103, 112
  transmission, 97, 98, 107
  unmodulated chirps, 103
  upchirps, 97
Propagation loss estimation models
  additional losses
    scattering obstacles, 70, 72, 73
  building penetration losses
    3GPP model, 73, 74
    WINNER+ and ITU-R models, 73–77
  channel centre frequency, 61
  extended 3GPP, 63
  Hata, 61, 62
  WINNER+, 63–65
PSD boosting, 162, 167, 172, 173

R
Radio Resource Control (RRC), 158
Random Access Channel (RACH), 177
Random Access Response (RAR), 153
Random Frequency and Time Division Multiple Access (RFTDMA)
  advantages, 92
  asynchronous access, 91
  BPSK demodulator, 92
  BW spectral range, 92
  competition protocol, 91
  frequency domain, 91
  multiple access mechanism, 92
  time domain, 91
  ultra-wideband systems, 91
Regularity, 138
Reverberation chamber, 131–135
Round Trip Time (RTT), 31

S
Short DRX cycle, 159, 160
Short-range devices (SRD), 56
SigFox system
  ALOHA protocol, 115
  communication protocol, 147
  data packet, 115
  end users, 114
  energy efficiency, 115
  features, 116, 117
  immunity, multipath propagation
    design simplicity, 149
    OM, 148
    PER measurements, 148
    system performance, 149
    transmission inability, 148
  immunity, noise and interference
    AWGN, 147
    operational assumptions, 147, 148
  microprocessor platforms, 114
  operator networks, 114
  random physical channels, 147
  RFTDMA protocol, 115
  time-spectral structure, 115, 116
  UNB, 114
Signal generator, 123, 126, 127
Signal-to-noise-and-interference ratio (SNIR), 43, 79
Smart grid, 26
Smart metering, 30
  businesses and individual consumers, 33
  fleet management, 36
  issue, 34
  M2M structure, 34
  M2M systems, 36
  reporting systems
    ETSI TS 102689, 34
    ETSI TS 102690, 34
    TS 102921, 34
  standardization mandate, 33
  traditional, 33
Software as a Service (SaaS), 23
Stand-alone' scenario, 161
Statistical traffic model
  emergency traffic, 31
  EU, 31
  periodic reporting, 32
  triggered reporting, 31
Stirrer, 133–135
Suburban, 62, 72
SyncWord, 101, 103
System Frame Number (SFN), 158
System Information Block (SIB2), 158

T
Tactile Internet, 22
Telecommunication traffic, 25
Time Division Multiplexing (TDM), 173
Time-hopping spread spectrum (THSS), 81
Transmit Block Size (TBS), 171
Two-way communication, 36

U
Ultra-narrowband (UNB), 51
Ultra-narrowband cooperative-UNB (C-UNB), 153
Ultra-Narrowband Transmission (UNB) systems
definition, 87
DL radio interface, 90
FIR, 87
nonoverlapping channels, 88
radio channel, 88
RFTDMA, 88, 91
transmission, 87, 88
UL radio interface, 88, 89
Uplink (UL) radio interface, 88, 89
Uplink traffic (UL), 112
User equipment (UE), 42, 156
User terminal (UE), 151

V
Vehicle-to-Infrastructure (V2I), 36
Virtual reality (VR), 22

W
Weightless (-P) system, 112
data packet, 112
ERP and DC, 110
FSK modulation, 112
GMSK modulation mode, 112
immunity, multipath propagation
individual stirrer positions, 145
long-distance communication, 146
NB and WB modes, 145
OM, 144
PER response dynamics ranges, 146
PER test, 144
radio channel change, 145
recommendation, 146
static and dynamic modes, 144
total transmission disruption, 146
immunity, noise and interference
AWGN signal, 142
CNR requirements, 143
difference analysis, 142, 143
dynamic range, 144
NB and WB, 144
OM, 142
parameters, 142
wideband, 142
IoT space, 111, 112
ISM frequency bands, 110, 111
N standard, 113
NB and WB, 112
open communication, 110
spectrum spreading, 110, 111
spread spectrum transmission, 110
UL and DL, 112
W standard, 113
Weightless SIG consortium, 112
Weightless-N standard, 113
Weightless-N system, 53
Weightless-W standard, 113
White TV spaces, 113
Wideband (WB), 112, 144
WINNER+ model, 63–65
Wireless M-Bus, 40
Wireless sensor network (WSN), 17
WUST ECL reverberation chamber, 148

Z
ZigBee system, 86

The manufacturer's authorised representative in the EU is Springer Nature Customer Service Centre GmbH, Europaplatz 3, 69115 Heidelberg, Germany. If you have any concerns regarding our products, please contact ProductSafety@springernature.com

Printed and bound by CPI Group (UK) Ltd, Croydon, CR0 4YY
05/10/2026
02228124-0003